Lecture Notes in Computer Science   7781

Commenced Publication in 1973
Founding and Former Series Editors:
Gerhard Goos, Juris Hartmanis, and Jan van Leeuwen

Jan Jürjens   Benjamin Livshits
Riccardo Scandariato (Eds.)

Engineering Secure Software and Systems

5th International Symposium, ESSoS 2013
Paris, France, February 27 – March 1, 2013
Proceedings

 Springer

Volume Editors

Jan Jürjens
TU Dortmund and Fraunhofer ISST
Department of Computer Science
Otto-Hahn-Straße 14
44221 Dortmund, Germany
E-mail: jan.jurjens@cs.tu-dortmund.de

Benjamin Livshits
Microsoft Research
One Microsoft Way
98052-6399 Redmond, WA, USA
E-mail: livshits@microsoft.com

Riccardo Scandariato
KU Leuven
Department of Computer Science
Celestijnenlaan 200A
3001 Heverlee, Belgium
E-mail: riccardo.scandariato@cs.kuleuven.be

ISSN 0302-9743   e-ISSN 1611-3349
ISBN 978-3-642-36562-1   e-ISBN 978-3-642-36563-8
DOI 10.1007/978-3-642-36563-8
Springer Heidelberg Dordrecht London New York

Library of Congress Control Number: 2013930843

CR Subject Classification (1998): E.3, D.4.6, D.2.1, D.2.4, F.3.1, K.6.5

LNCS Sublibrary: SL 4 – Security and Cryptology

Typesetting: Camera-ready by author, data conversion by Scientific Publishing Services, Chennai, India

Printed on acid-free paper

Springer is part of Springer Science+Business Media (www.springer.com)

Preface

It is our pleasure to welcome you to the fifth International Symposium on Engineering Secure Software and Systems. This event in a maturing series of symposia attempts to bridge the gap between the scientific communities from software engineering and security with the goal of supporting secure software development. The parallel technical sponsorship from ACM SIGSAC (the ACM interest group in security) and ACM SIGSOFT (the ACM interest group in software engineering) demonstrates the support from both communities and the need for providing such a bridge.

Security mechanisms and the act of software development usually go hand in hand. It is generally not enough to ensure correct functioning of the security mechanisms used. They cannot be "blindly" inserted into a security-critical system, but the overall system development must take security aspects into account in a coherent way. Building trustworthy components does not suffice, since the interconnections and interactions of components play a significant role in trustworthiness. Lastly, while functional requirements are generally analyzed carefully in systems development, security considerations often arise after the fact. Adding security as an afterthought, however, often leads to problems. Ad hoc development can lead to the deployment of systems that do not satisfy important security requirements. Thus, a sound methodology supporting secure systems development is needed.

The conference program includes two major keynotes from Laurie Williams (NC State University) on why we need a science for software security, and George Danezis (Microsoft Research) on privacy-enhancing technologies, as well as a set of research and idea papers. In response to the call for papers, 62 papers were submitted. The Program Committee selected 15 contributions (24%), presenting new research results on engineering secure software and systems. These include two idea papers, giving a concise account of new ideas in the early stages of research.

Many individuals and organizations have contributed to the success of this event. First of all, we would like to express our appreciation to the authors of the submitted papers and to the Program Committee members and external referees, who provided timely and relevant reviews. Many thanks go to the Steering Committee for supporting this series of symposia, and to all the members of the Organizing Committee for their tremendous work and for excelling in their respective tasks. The DistriNet research group of the KU Leuven did an excellent job with the website and the advertising for the conference. Finally, we owe gratitude to ACM SIGSAC/SIGSOFT, IEEE TCSP and LNCS for continuing to support us in this series of symposia.

December 2012

Jan Jürjens
Benjamin Livshits
Riccardo Scandariato

Conference Organization

General Chair

Valérie Issarny   Inria Paris-Rocquencourt, France

Program Co-chairs

Jan Jürjens   TU Dortmund and Fraunhofer ISST, Germany
Ben Livshits   Microsoft Research in Redmond, USA

Publication Chair

Riccardo Scandariato   KU Leuven, Belgium

Publicity Chair

Pieter Philippaerts   KU Leuven, Belgium

Web Chair

Ghita Saevels   KU Leuven, Belgium

Local Arrangements Co-chairs

Chantal Girodon   Inria Paris-Rocquencourt, France
Stéphanie Chaix   Inria Paris-Rocquencourt, France
Emmanuelle Grousset   Inria Paris-Rocquencourt, France
Florence Barbara   Inria Paris-Rocquencourt, France

Steering Committee

Jorge Cuellar   Siemens AG, Germany
Wouter Joosen   Katholieke Universiteit Leuven, Belgium
Fabio Massacci   Università di Trento, Italy
Gary McGraw   Cigital, USA
Bashar Nuseibeh   The Open University, UK
Daniel Wallach   Rice University, USA

Program Committee

Davide Balzarotti	EURECOM, France
Ruth Breu	University of Innsbruck, Austria
Cristian Cadar	Imperial College, UK
Julian Dolby	IBM Research, USA
Matt Fredrikson	University of Wisconsin, USA
Dieter Gollmann	TU Hamburg-Harburg, Germany
Maritta Heisel	U. Duisburg Essen, Germany
Peter Herrmann	NTNU, Trondheim, Norway
Thorsten Holz	U. Ruhr Bochum, Germany
Sergio Maffeis	Imperial College, UK
Heiko Mantel	TU Darmstadt, Germany
Anders Møller	Aarhus University, Denmark
Haris Mouratidis	University of East London, UK
Zachary Peterson	Naval Postgraduate School, USA
Frank Piessens	KU Leuven, Belgium
Erik Poll	RU Nijmegen, The Netherlands
Alexander Pretschner	TU Munich, Germany
Wolfgang Reif	University of Augsburg, Germany
Jianying Zhou	Institute for Infocomm Research, Singapore
Mohammad Zulkernine	Queens University, Canada

Additional Reviewers

Pieter Agten	Petr Hosek	Giancarlo Pellegrino
Azadeh Alebrahim	Jinwei Hu	Matthias Perner
Andreas Angerer	Bart Jacobs	Alfredo Pironti
Kristian Beckers	Kuzman Katkalov	Rodrigo Roman
Leyla Bilge	Florian Kelbert	Thomas Santen
Matthias Büchler	Abdullah Abdul Khadir	Bagus Santoso
Aldar Chan	Frank A. Kraemer	Jens Sauer
Antoine Delignat-Lavaud	Prachi Kumari	Hella Seebach
Stephan Faßbender	Enrico Lovat	Florian Siefert
Daniele Filaretti	Alexander Lux	Christian Sillaber
Richard Gay	Aderhold Markus	Jan Smans
Sylvia Grewe	Rene Meis	Ben Smyth
Linda Ariani Gunawan	Rabih Mohsen	Kurt Stenzel
Axel Habermaier	Jan Tobias Muehlberg	Raoul Strackx
Dina Hadziosmanovic	Florian Nafz	Thomas Trojer
Jin Han	Nick Nikiforakis	Dries Vanoverberghe
Denis Hatebur	Johan Oudinet	K. Weldemariam
Dominik Holling	Hristina Palikareva	Philipp Zech

Sponsoring Institutions

Inria, France

NESSoS FP7 Project, Network of Excellence
on Engineering Secure Future Internet Software
Services and Systems, www.nessos-project.eu

Table of Contents

Formal Methods

Analyzing

Control-Flow Integrity in Web Applications

Bastian Braun, Patrick Gemein, Hans P. Reiser, and Joachim Posegga

Institute of IT-Security and Security Law (ISL), University of Passau
{bb,hr,jp}@sec.uni-passau.de, pgemein@gmx.de

Abstract. Modern web applications frequently implement complex control flows, which require the users to perform actions in a given order. Users interact with a web application by sending HTTP requests with parameters and in response receive web pages with hyperlinks that indicate the expected next actions. If a web application takes for granted that the user sends only those expected requests and parameters, malicious users can exploit this assumption by crafting harming requests. We analyze recent attacks on web applications with respect to user-defined requests and identify their root cause in the missing explicit control-flow definition and enforcement. Based on this result, we provide our approach, a control-flow monitor that is applicable to legacy as well as newly developed web applications. It expects a control-flow definition as input and provides guarantees to the web application concerning the sequence of incoming requests and carried parameters. It protects the web application against race condition exploits, a special case of control-flow integrity violation. Moreover, the control-flow monitor supports modern browser features like multi-tabbing and back button usage. We evaluate our approach and show that it induces a negligible overhead.

1 Introduction

Over the past two decades, the web has evolved from a simple delivery mechanism for static content to an environment for powerful distributed applications. In spite of these advances, remote interactions between users and web applications are still handled using the stateless HTTP protocol, which has no protocol level session concept. Handling session state is fully left to the web application developer or to high-level web application frameworks.

Web applications often include complex control flows that span a series of multiple distributed interactions. The application developer usually expects the user to follow the intended control flow. However, if a web application does not carefully ensure that interactions adhere to the intended control flow, attackers can easily abuse the web application by using unexpected interactions. Several known attacks have exploited this kind of vulnerability in the past. The attacks' impact ranges from sending more free SMS text messages than actually allowed [1], over unauthorized access to user accounts [2,3,4], up to shopping expensive goods with arbitrarily low payments [5].

This paper presents novel approaches for avoiding problems related to control-flow integrity in web applications. The specific contributions are as follows.

J. Jürjens, B. Livshits, and R. Scandariato (Eds.): ESSoS 2013, LNCS 7781, pp. 1–16, 2013.
© Springer-Verlag Berlin Heidelberg 2013

First, we define a formal language for specifying explicitly the control flow of a web application; Second, we define a control mechanism that makes sure that only client requests that comply with the control-flow specification are executed; Third, we integrate the control mechanism in a framework based on the Model-View-Controller (MVC) model, making our approach both easy to use for newly developed applications and easy to integrate in already existing applications. Finally, we show that our approach is effective and practical by demonstrating that it enables the removal of several kinds of real-world security problems, while having a low run-time overhead.

This paper is structured as follows. The next section provides an in-depth discussion of technical aspects of control-flow integrity in web applications and explains known attacks and vulnerabilities. Section 3 presents novel approaches for controlling flow integrity at the server-side. Section 4 analyzes the benefits of our approach and evaluates the performance of our prototype. Section 5 compares our approach to related work, and Section 6 concludes.

2   Web Application Control Flow

In this section, we investigate in more detail the problem of control-flow integrity of web applications, analyze several real-world attacks, and discuss their root causes.

2.1   Technical Background

In a typical web application, the user's web browser interacts with the remote application by sending HTTP requests. HTTP is a stateless protocol without session concept [6]. This means that each request is independent of all others. The protocol does not inherently link one request to the next. Dynamic web applications, however, have workflows that are composed of multiple steps, which corresponds to multiple HTTP requests from the user to the web application. For each step, the client receives a web page with hyperlinks that offer possible next steps to a user. Upon clicking a link, the user's browser sends a particular HTTP request to the web application, which then performs actions in order to progress to the next step in the workflow. The actions are defined by the URI [7] of the HTTP request, the request parameters, and the server-side session record.

2.2   The Attacks

Several kinds of attacks exploit the fact that attackers can craft arbitrary requests instead of clicking on provided hyperlinks. Real-world examples of control-flow integrity violations are race conditions, manipulated HTTP parameters, unsolicited request sequences, and the compromising use of the browser's "back" button.

Race Conditions. In order to exploit race conditions [8] in web applications, attackers can send several crafted requests almost in parallel. If the web application does not handle concurrent requests by proper synchronisation, the actual application semantics can be changed in this way. In one real-world example, a web application provided an interface to send a limited number of SMS text messages per day [1]. The web application first checked the current amount of sent messages (*time-of-check*), then delivered the message according to the received request, and finally updated the number of sent messages in the database (*time-of-use*). Attackers were able to send more messages than allowed by the web application by crafting a number of HTTP requests, each containing the receiver and text of the message to be sent. These requests were sent almost in parallel and the multi-threaded web application processed the incoming requests concurrently. This way, the attacker exploited the fact that the messages were sent before the respective database entry was updated, leading to the delivery of all requested messages. The developers' underlying assumption was that users finish one transmission process before sending the next message and do not request one operation of the workflow several times in parallel.

HTTP Parameter Manipulation. HTTP requests can contain parameters in addition to the receiving host, path, and resource. As the parameters are sent by the client, the user can control the parameters' values and which parameters are sent to the web application. Wang et al. [5] found a bunch of logic flaws in well-known merchant systems and Cashier-as-a-Service (CaaS) services. These flaws allowed them to buy any item for the price of the cheapest item in the store. In 2011, the Citigroup faced an attack on their customers' data [4]. The attackers were able to access names, credit card numbers, e-mail addresses and transaction histories. All the attackers had to do was simply changing the HTTP parameters in the web browser. By automation, they obtained confidential data of more than 200,000 customers.

Unsolicited Request Sequences. Attackers can not only modify the requests' parameters but also craft requests to any method of the web application. Besides manipulated HTTP parameters, web applications might face unexpected requests to any method. For instance, in another given scenario by Wang et al. [5], a malicious shopper was able to add items to her cart between checkout and payment. She was only charged the value of her cart at checkout time. The recently added items were not invoiced.

Compromising Use of the "Back" Button. Current web browsers are fitted with a so-called "back" button. It is meant to navigate back to the last visited web page. Depending on the configuration, the last request either has to be repeated in order to display the page or the content is loaded from the browser's local storage ("cache"). In the context of a workflow, the user takes one step back which in some cases is unwanted and also undetectable by the web application. In fact, the usage of this button usually invokes the last action again rather than rolling back the last changes. Hallé et al. [9] describe related navigation errors.

2.3   Root Causes

All described attacks share common root causes. Web application developers assume that users first request one of possibly several application entry points, e.g. the base directory at `http://www.example.com`. Upon the first request, the web application sends a given response containing a set of hyperlinks or a redirect instruction to the user. As users tend to click on hyperlinks in order to navigate through the application, developers might assume that only the given requests will be accessed next. However, the user is technically not bound to click on one of the provided hyperlinks but she can still send requests that are not provided within this response. Sent requests can differ from provided hyperlinks in terms of addressed methods and HTTP parameters. Vulnerable web applications fail to handle unintended user behavior in terms of sequences of requests.

More formally, web application developers implement implicit control-flow graphs. In each state, sending a request leads to a subsequent state in the graph. Executing a step corresponds to changing the server-side state. Control-flow weaknesses occur if an attacker is able to address at least one method, i.e. cause a state-changing action, that is not meant to be addressed in the respective session state. In the respective control-flow graph, this transition does not exist due to the developer's assumption that the request does not happen at that time. Vice versa, a web application implementing a control-flow graph with transitions for all requests in every state is not susceptible to control-flow weaknesses.

Control-flow weaknesses cannot be overcome with usual access control means. The attack vectors include only requests that are in the scope of the user's rights. Access control mechanisms prevent users from accessing sensitive API methods at all time while control-flow integrity protection prohibits access to regular API methods at the wrong time.

Existing web applications enforce the intended control flow based on session-contained parameters. This allows only the implicit definition of workflows. The previous actions are assumed to set the parameters and, thus, allow the execution of next actions. The actual workflows are not explicitly determined preventing the proper assessment of enabled workflows. The central and explicit definition of facilitated workflows provides guarantees of request sequences to the relying web application. One crucial aspect of reliable request sequences are controlled HTTP parameters as we have shown by the attacks in Section 2.2.

3   Preserving Control-Flow Integrity

The attacks described in Section 2.2 are caused by user actions that violate given control flows. This section provides detailed information of how we prevent an unintended action from getting executed and, thus, from violating the integrity of a control flow.

3.1   Technical Background

For every web application, the application developer knows the intended control flow. This control flow can be denoted as a sequence of actions. Considering each action as a transition in a graph, we finally obtain the control-flow graph of the web application. So, the application developer deploys the control-flow graph of the web application.

The enforcement of the intended control flow requires a central entity that takes care of each incoming request. The popular Model-View-Controller architecture provides such an entity by design (see Figure 1). Every request has to pass the application's controller, which encapsulates the business logic. The controller consists of several classes, each containing various methods. Therefore, one action of a control flow in our definition language is defined as <class name>.<method name>. From a granularity view, this is appropriate because a request addresses one method. In sum, a control-flow graph is given as a sequence of methods of controller classes.

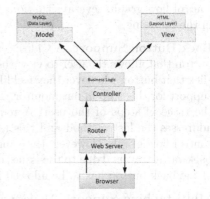

Fig. 1. The Design Pattern of MVC-based Web Applications

3.2 Protection Goals

Our approach protects web applications from malicious users that perform attacks using arbitrary request sequences. As a side effect, the approach protects honest users against *Cross-Site Request Forgery (CSRF)* attacks to some extent because attackers have to follow the intended control flow to finally commit their abusive request. In more detail, our approach has the following goals:

– Upon each incoming request, the monitor shall determine the control-flow context of this request and take a decision whether the request is permitted. If the monitor allows the request to pass, it updates the context accordingly.
– The approach must be usable with state-of-the-art browser features, including the use of a back button as well as multiple browser tabs (multi-tabbing) for the same session. Each tab shall be permitted to use a different control flow.
– The monitor must prevent race conditions for actions that might serve as a target for an attacker. These actions can be specified in the control-flow graph.
– The monitor must be able to control HTTP parameters and their values.
– All web applications have unclassified resources such as "About us" information. These resources shall be accessible without restrictions, independent of ongoing workflows.

3.3 Enforcing Control-Flow Integrity

In this section, we provide details how we achieve the above-mentioned goals. We propose an architecture based on explicit control-flow specification and server-side control-flow enforcement. The central control-flow monitor combines several mechanisms to enforce control-flow integrity. We show that all user interactions are intercepted and checked by our control-flow monitor. Besides simple checks that sequences of requests are compatible with sequences in the control-flow graph, several situations require dedicated treatment, as explained in the following.

Back Button Support. A widespread feature of modern browsers is the back button that allows the user to view the last web page again. As users are used to click that button whenever they feel like revisiting the last page, we implemented support for this step in our monitor. Therefore, the control-flow monitor records the trace of steps of the user. A request is considered a step backwards if it addresses the last method and this method is not meant as a next step in the control-flow graph. However, the control-flow monitor by default prohibits the backwards traversal due to the issues described in Section 2.2. Instead, the usage of the back button has to be allowed in the control-flow graph for each step.

Multi-tabbing Support. Modern web browsers usually allow several tabs in the same window. As these tabs share the client-side data, e.g. cookies [10], across all instances, they are hardly distinguishable from the server side. Hence, without multi-tabbing support, actions in one tab would violate the control flow in another. In order to overcome this drawback, the control-flow monitor inserts client-side identifiers for different tabs to tell them apart. This way, each tab can be treated individually though logged in at the same web application.

Race Condition Prevention. The monitor prevents the exploitation of race condition vulnerabilities (see Section 2.2), by disabling parallel execution of susceptible actions. In general, these are actions that add, update, or delete data after reading. We achieve this goal with a locking mechanism. The control-flow monitor creates a temporary lock named by the session ID of the user. This means race condition protection on session level. Moreover, protection on control-flow and user level is possible by using a control-flow ID and the user ID respectively. Even a system-wide protection can be implemented using one unique ID file for all users.

Parameter Validation. The client-side manipulation of HTTP parameters can lead to unintended application states (see Section 2.2). Thus, request parameters have to be checked for validity on the server side before they are processed. Instead of leaving this task to each method, the control-flow monitor provides means to centrally enforce given parameter properties. First, the data type of each parameter can be defined. As a side effect, this feature also mitigates *injection attacks (XSS, SQLi)* that need to transmit control characters. Second, parameters can be marked as "write once read many" (WORM). This allows to set the parameter's value once but not change it afterwards, meaning that this value is immutable for the rest of the session. This provides an invariant guarantee to the web application. One use case is the user ID that is supposed to not change during a session. Third, parameter names can be excluded for given workflows. This feature can protect web applications from unintended data manipulation. For instance, it prevents the setting of control flow-invariant parameters.

Definition of Uncritical Methods. All web applications contain uncritical methods. Accessing these methods does not harm the application's control-flow integrity. For instance, a chat function can be allowed beside the enforced workflow. Similarly, AJAX calls that update the user's view but do not change the application's state can also be allowed.

Control-Flow Definition. In this section, we provide details on the syntax of the control-flow graph definition language. The following clauses and operators can be combined recursively.

`Method1` → `Method2` — After accessing `Method1`, the user is allowed to access `Method2`.
`(Method1|Method2)` — The user is allowed to access `Method1` or `Method2` in the first place, but she is not allowed to change her decision after clicking the back button.
`(&Method1|&Method2)` — Like above but the user is allowed to change her decision after clicking the back button – denoted by the `&` symbol.
`@Method{x}` — The user is allowed to access `Method` repeatedly. It is possible to define a maximum number `x` of allowed executions.
`?Method` — The back button support for `Method` is enabled, i.e. the user can navigate one step backwards after having called this method.
`!Method` — The race condition protection is active for this method. As long as this method is executed, no other protected method is executed in the context of the same session, user account, or system-wide (see above).
`Method[+par1=type1,*par2=type2]` — Only parameters `par1` and `par2` are allowed for `Method` where they can be sent via POST (`+`) or GET (`*`) and have data types `type1` and `type2` respectively. Predefined data types include `bool`, `numeric`, and `string`. A policy for the whole control flow can be set by `addParameterTypeGlobal("*par=type")`.
`addForbiddenParameters("par")` — Parameter `par` must not occur in the whole control flow.
`addParametersGlobal("par")` — Parameter `par` can be set once but is immutable afterwards.

The nesting of clauses allows for defining complex control-flow policies. We provide simple examples in Section 3.5 and a more sophisticated case in Section 4.

3.4   The Implementation

For implementing our control flow monitor, several challenges need to be addressed. Most importantly, the monitor has to be integrated into an application framework, which can be a complex task especially for existing applications. In addition, handling race conditions and multitabbing also deserve more detailed attention.

Integration into Web Applications. We implemented our control-flow monitor as a PHP module. It is run by the router (see Figure 2) before the controller class is called. This strategic position makes sure that, first, all requests have to pass our control-flow monitor before being processed by the web application and, second, the monitor is easy to integrate into existing web applications.

As a proof of concept, we integrated the monitor into a web application that is based on the CodeIgniter

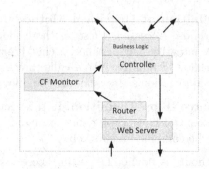

Fig. 2. Implemented Modification of the Design Pattern of MVC-based Web Applications w.r.t. Figure 1

framework [11]. In fact, the only change on an existing web application affects the one line of code that calls the responsible controller. This line has to be slightly modified to include our monitor (see Listing 1.1).

```
include(APPPATH.'controllers/'.$RTR->fetch_directory().$RTR
    ->fetch_class().'.php');
//must be changed to
AOP::process(APPPATH.'controllers/'.$RTR->fetch_directory().
    $RTR->fetch_class().'.php',
    $_SESSION[''atom_parentFramework'']->getCacheFolderName
        ());
```

Listing 1.1. Dynamic Inclusion of Controller Classes in the CodeIgniter Framework [11]

We use *Aspect Oriented Programming (AOP)* to inject the control-flow monitor as a processing step into the call sequence of all controllers. This allows the developer to apply changes on the application the same way as if there were no control-flow monitor.

Multi-tabbing Support. As explained before, multi-tabbing requires the unique identification of tabs. This identification is implemented in JavaScript. Moreover, a tab handler is implemented on server side as part of the control-flow monitor. The JavaScript code triggers an AJAX message whenever a tab is opened, closed, or a tab switch is performed by the user. A tab switch message by the client makes the tab handler change the tab context on server side. When the user opens a new tab by clicking on the "open link in new tab" option in the browser, this tab is assigned a session-unique identifier. We use the `window.name` property of the `window` DOM object to store the identifier. An AJAX request transmits the new identifier to the tab handler. The new tab is assigned the advanced position in the control-flow graph while the first tab holds the former position. Both tabs then run the same control flow, however, it is enforced individually, i.e. a control-flow violation in one tab has no effect on the other tab as the respective tab record is duplicated when the new tab is opened.

The control-flow monitor stores flow-related information per tab, i.e. the active control-flow graph that is currently enforced in this tab and the respective position in the graph. The user's session ID and other high level information is still stored in the session record. This allows, for instance, to consider several products in different tabs, then add some of them in the same shopping cart and finally check out in one tab that starts the checkout control flow.

An attacker stripping or manipulating the embedded tracking code can not trick the system to gain advantages. The code only signals the current tab to the web application. A manipulation would cause the web application to assign the next request to a different tab. This, however, is equivalent to perform the request in the respective tab. The intended action is only executed if the request is allowed there. Then, however, the attacker has not increased his scope of action. In all other cases, the manipulation leads to voiding the current control flow.

Race Condition Prevention. Whenever a protected method is executed, the control-flow monitor tries to create a file with the current session ID. If this creation fails due to an existing file with the same name, the request is not processed and an error page is displayed. After processing the protected method, the file lock is released again. This allows the next protected method to be executed.

The race condition protection mechanism does not prevent the processing of unprotected methods, e.g. in a different tab. The fine granularity of the locking makes sure that a single locked method has no impact on the usability of other sessions of the user or the interactions of other users.

3.5 Simple Examples

In this section, we show the usage of our control-flow definition language. We give examples with respect to the real-world scenarios in Section 2.2 but assume a simplified technical implementation to keep the control flows simple and clear. We give details on the application of our control-flow monitor in the context of the Amazon checkout process in Section 4.

Preventing Race Conditions in SMS Delivery. In the first example [1], attackers managed to bypass the delivery limit of an SMS portal by exploiting a race condition vulnerability. We assume the following control flow to send an SMS: First, the user requests the SMS input form. Then, after entering all necessary information, the user submits the form. The related control-flow definition ensures that, first, the input form has to be accessed before the submission, and, second, the submission must be protected against race condition attacks, see Listing 1.2.

```
SMS.showForm -> !SMS.validateAndSendForm
```

Listing 1.2. Control-Flow Definition to Protect SMS Delivery from Race Conditions

The control-flow definition allows access to the method `validateAndSendForm` only after requesting `showForm`. This prevents the attacker from sending the message information directly to the delivery gateway. Of course, a capable attacker might send the requests to the `showForm` method in an automated fashion. However, as the `validateAndSendForm` method is protected against race condition attempts, e.g. on user level, the attacker's requests will only be processed sequently. This avoids sending more messages than actually allowed.

Prevent Adding Items to the Shopping Cart between Checkout and Payment. A more complex example is given by Wang et al. [5]. After requesting the checkout, the user was able to add more items to her shopping cart. These items were not charged. In order to prevent this sequence of requests, the checkout workflow has to be properly defined. The method that adds goods to the cart must not be accessible during this workflow. The respective control flow definition is given in Listing 1.3.

```
Checkout.logIn
-> Payment.chooseMethod
-> Payment.validateStatus
-> Checkout.completeOrder
```

Listing 1.3. Control-Flow Definition to Prevent Adding Items after Checkout

After the authentication, i.e. login, the user chooses her favorite payment method and is redirected to a payment service provider. The actions on the payment service provider's site are not part of the definition because they happen on a different domain that is not controlled by the same control-flow monitor. The next request within the scope of the definition is the payment status validation after the user's return. Finally, the order is completed, the goods are shipped to the user, and the cart is reset. During the whole process, no addition of items to the cart is granted.

4   Discussion and Evaluation

In this section, we discuss the properties of our control-flow monitor. We show that it produces a negligible overhead and evaluate the protection goals defined in Section 3.2. Finally, we explain its possible application scenarios and limits.

4.1   Performance Evaluation

As described in Section 3.4, the control-flow monitor is applied between the router and the controller of the web application. It examines the received HTTP request with respect to the requested method and the parameters and checks these against a given policy. This application overhead is independent of the web application's execution time. The delay relates to the complexity of the given policy, though.

We used Xdebug (version 2.1.2) [12] to determine the control-flow monitor's overhead in a virtual machine with Debian 6 as the operating system and Apache2 as the web server with PHP 5.5.3.3-7 on an Intel Core-i7-2600 (Intel-VT activated) with 3.4 GHz and 2 GB RAM. For evaluation purposes, we implemented the checkout process of Amazon. Therefore, we analyzed the control flow on amazon.com by hand and derived the control-flow definition given in Listing 1.4. Note that the controller and method names are simplified for readability reasons. The control-flow definition does not allow usage of the back button because Amazon prohibits it, too.

```
1    Login.index
2    -> (Address.chooseExisting | Address.addNew)
3    -> Shipping.preferences
4    -> ((Payment.chooseExisting
     | Payment.addNewCreditCard)
     | Payment.addNewDebitCard)
5    -> (Billing.chooseExisting | Billing.addNew)
6    -> Order.placeOrder
```

Listing 1.4. Definition of Amazon's Checkout Control Flow

Table 1. Overhead caused by the control flow monitor in ms

Step	Runs										
	1st	2nd	3rd	4th	5th	6th	7th	8th	9th	10th	avg
1	8.9	8.2	8.4	10.2	11.0	8.7	8.2	9.4	8.3	7.7	8.9
2	10.2	9.9	9.3	9.8	10.1	9.8	9.0	8.2	9.5	9.1	9.5
3	10.1	9.2	10.9	8.6	9.6	8.2	9.5	9.0	9.0	8.4	9.2
4	8.3	10.1	10.0	10.2	9.4	9.8	10.3	7.8	10.6	9.0	9.6
5	8.8	11.0	10.1	8.3	8.4	10.0	8.6	7.9	7.7	9.8	9.1
6	10.0	8.5	8.1	8.5	8.4	8.4	9.6	9.7	8.0	10.4	9.0

We measured the runtime overhead ten times and computed the average for each step in the control flow (see Table 1). The respective graph shows a peak in the fourth state due to the triple branching (see Figure 3). Branches, namely alternative paths through the control flow, cause most of the overhead, the earlier a branch occurs the bigger its overhead. This is the reason why step 2 causes more overhead than step 5. We assume that

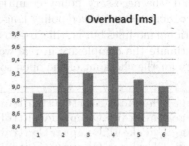

Fig. 3. Performance Evaluation of the Amazon Checkout Process

some overhead can be saved by a more efficient policy parsing algorithm. Overall, the induced delay ranges between 8.9 and 9.6 milliseconds per request. We consider this an acceptable effort with respect to the security gain. In order to determine the monitor's scalability to several user sessions, we set up 100 parallel user sessions and repeated the measurement. While the overall response time increased, we found out that there is no measurable difference to the scenario with only one user in terms of the monitor's overhead.

There is a one-time overhead for the generation of the temporary controller class file (see Section 3.4). This overhead occurs once whenever a new controller class is added or an existing class is modified. The first call on this class takes 60% to 90% more time than the subsequent calls. For usability concerns, this overhead can be neglected because the web application provider could easily initiate an appropriate request, thus, preventing all users from facing the mentioned delay.

4.2   Discussion

In this section, we evaluate our findings with respect to the protective goals defined in Section 3.2. We have to note that the monitor is responsible for control-flow integrity while other tasks like session management and user authentication are handled by the framework in place.

Every incoming HTTP request has to pass the router in the assumed MVC architecture. So, all requests are finally processed by our control-flow monitor. Our security evaluation showed that in fact all requests are treated by the monitor and accepted or rejected appropriately. The control-flow monitor achieves complete protection against maliciously crafted requests as well as erroneous navigation attempts.

However, the protection level depends on the sound definition of control flows. The definition has to be provided by the application developer. The implications from this fact are twofold. First, the definition requires a deep knowledge of the web application and its methods. The knowledge and understanding of the web application must already exist to implement and maintain the web application. This allows developers to provide more accurate control flow policies than automatic approaches. So, we consider this a feasible task for an expert. Second, the necessary policy definition efforts stay within reasonable bounds. The class.method-based policy language abstracts from the implementation of functional modules but is still close to the web application's architecture. In order to estimate the complexity in real world use cases, we crafted the control flow policy for the checkout of the open source shop Magento [13] (see Listing 1.5).

```
Mage_Checkout_CartController.indexAction ->
Mage_Checkout_OnepageController.indexAction ->
Mage_Checkout_OnepageController.saveBillingAction ->
Mage_Checkout_OnepageController.saveShippingAction ->
Mage_Checkout_OnepageController.saveShippingMethodAction ->
Mage_Checkout_OnepageController.savePaymentAction ->
Mage_Checkout_OnepageController.reviewAction ->
Mage_Checkout_OnepageController.successAction
```

Listing 1.5. Control-Flow Definition of the Magento [13] Online Shop

Our control-flow monitor provides multi-tabbing and back button support, thus proves usable with modern browser features. This increases usability and ensures acceptance by end users. This way, security is not achieved at the expense of a limited user experience.

To the best of our knowledge, our approach is the first to effectively protect against race condition exploits. The control-flow monitor allows the flexible definition of the protection level, ranging from control flow-based over user-level up to system-wide protection.

Policies on HTTP parameters can be defined including both GET and POST parameters. Policy rules can apply in terms of data type, the limitation on a single value assignment, and the exclusion of parameters for given workflows. Our parameter control means are suitable to prevent the attacks described in Section 2.2.

The definition of uncritical methods allows control-flow integrity to focus on a comprehensible set of relevant method calls. For instance, there can be unhindered access to pictures because they are not part of the business logic. Confidential data can be protected by access control means. AJAX requests can be divided into state-changing and other requests. The state-changing requests can be covered by the control-flow definition, the others are excluded and pass the control-flow monitor. As AJAX requests also call server-side methods, their control-flow definition is straightforward with respect to the web application's control flow.

Our approach is easily applicable at development phase though one of its most advantageous features is its usability with legacy web applications. We implemented a PHP-based proof of concept. Nevertheless, a Java-based implementation can be achieved with acceptable effort, e.g. by a J2EE filter. Even non-MVC-based web applications can be equipped with the monitor. However, the integration causes more overhead if a central request handler is missing. Then, all calls on server-side actions have to be intercepted separately.

The control-flow monitor does not aim at replacing the web application's business logic. As a matter of fact, it provides reasonable and reliable guarantees concerning the sequence of requests and properties of provided parameters. The web application still has to make sure that user generated content fits the expected information. For instance, a sequence of requests containing semantic garbage but matching the defined control flow will still succeed to finally request the intended method.

5   Related Work

The *Open Web Application Security Project (OWASP)* coined the term *Failure to Restrict URL Access* [14] to describe a similar vulnerability as our control-flow weakness. However, it is more focused on access control flaws that can be exploited by *Forced Browsing attacks* [15] to find a *deep link* [16] to a high privilege web page. Workflows and control-flow integrity play a tangential role in the description.

To the best of our knowledge, there is no work with a similar scope of protection and a comparable feature set. The approaches that restrict the web application's request surface towards the user are considered most similar to our approach. They limit the accepted requests to a predefined set and prevent arbitrary navigation by users. This is either achieved by issuing tickets to access server-side resources [17] which, by design, inhibits multitabbing and back button support as well as page reloads, or by defining pairs of steps that can be executed in order [9] where the first step serves as a gatekeeper to the second step. The latter approach does not allow to define complete workflows explicitly what we identified as a crucial point.

Other approaches aim at detecting unintended or unusual server states [18], combinations of such server states and code execution points [19] or code execution paths that lead to the violation of application invariants [20] or input/output invariants [21]. These approches infer the intended application states during a training phase or by static code analysis. They do not intend to make workflows explicit and control the interactions with users.

Malicious users not only craft individual HTTP requests or manipulate request headers to achieve their goals. Depending on the business logic of the web application, changes on the client-side JavaScript code can cause damage to the application provider. Existing approaches replicate client-side computation on server-side to detect deviations [22,23], statically analyze JavaScript to determine the expected sequence of requests [24], or check the web application for exploitable HTTP parameter pollution vulnerabilities [25].

An attacker exploiting a race condition vulnerability [8] can execute one function more often than intended by the application developer. Paleari et al. [1] describe an approach to detect race condition vulnerabilities in web applications.

6   Conclusion

We explained the complex problem of control-flow vulnerabilities and showed its high practical relevance by real-world examples, i.e. existing vulnerabilities and attacks. We identified the root causes in the modular addressability of web applications together with the implicit and scattered definition of workflows. Our solution overcomes this problem by the explicit definition and enforcement of intended workflows. To the best of our knowledge, it is the first approach that covers the whole bandwidth of related vulnerabilities, including race conditions, HTTP parameter manipulation, unsolicited request sequences, and the compromising use of the back button. Moreover, it is the first approach that properly handles client-side features like back button usage and multitabbing. We showed that this approach can prevent all described attacks and causes negligible overhead.

In sum, we provided a thorough approach that is applicable to existing and newly developed web applications and provides guarantees to the developer concerning the sequences of incoming requests as well as the format and values of parameters. This allows to separate web application semantics and control-flow integrity. As a side effect, the presented approach mitigates *Cross-Site Request Forgery (CSRF)* and *injection (XSS, SQLi)* attacks.

Acknowledgments. This work was in parts supported by the EU Project WebSand (FP7-256964), `https://www.websand.eu`. The support is gratefully acknowledged.

References

1. Paleari, R., Marrone, D., Bruschi, D., Monga, M.: On Race Vulnerabilities in Web Applications. In: Zamboni, D. (ed.) DIMVA 2008. LNCS, vol. 5137, pp. 126–142. Springer, Heidelberg (2008)
2. Chen, S.: Session Puzzles - Indirect Application Attack Vectors (White Paper) (May 23, 2012),
   `http://puzzlemall.googlecode.com/files/Session%20Puzzles%20-`
   `%20Indirect%20Application%20Attack%20Vectors%20-`
   `%20May%202011%20-%20Whitepaper.pdf`
3. Grossman, J.: Seven Business Logic Flaws That Put Your Website At Risk (White Paper) (May 19, 2012),
   `https://www.whitehatsec.com/assets/WP_bizlogic092407.pdf`
4. The New York Times: Thieves Found Citigroup Site an Easy Entry (May 24, 2012),
   `http://www.nytimes.com/2011/06/14/technology/14security.html`
5. Wang, R., Chen, S., Wang, X., Qadeer, S.: How to Shop for Free Online – Security Analysis of Cashier-as-a-Service Based Web Stores. In: IEEE Symposium on Security and Privacy (2011)
6. Fielding, R., Gettys, J., Mogul, J., Frystyk, H., Masinter, L., Leach, P., Berners-Lee, T.: Hypertext Transfer Protocol – HTTP/1.1. RFC 2616 (June 1999),
   `http://www.w3.org/Protocols/rfc2616/rfc2616.html`
7. Berners-Lee, T., Fielding, R., Irvine, U., Masinter, L.: Uniform Resource Identifiers (URI): Generic Syntax. RFC 2396 (August 1998),
   `http://www.ietf.org/rfc/rfc2396.txt`
8. OWASP: Race Conditions (May 23, 2012),
   `https://www.owasp.org/index.php/Race_Conditions`
9. Hallé, S., Ettema, T., Bunch, C., Bultan, T.: Eliminating Navigation Errors in Web Applications via Model Checking and Runtime Enforcement of Navigation State Machines. In: ASE (2010)
10. Kristol, D., Montulli, L.: HTTP State Management Mechanism. RFC 2109 (February 1997), `http://www.ietf.org/rfc/rfc2109.txt`
11. ExpressionEngine Dev Team: CodeIgniter - Open Source PHP Web Application Framework (May 24, 2012), `http://www.codeigniter.com/`
12. Xdebug (June 05, 2012), `http://xdebug.org/`
13. Magento Commerce (September 24, 2012), `http://demo.magentocommerce.com/`
14. OWASP: Failure to Restrict URL Access (May 11, 2012), `https://`
    `www.owasp.org/index.php/Top_10_2010-A8-Failure_to_Restrict_URL_Access`
15. OWASP: Forced Browsing (May 04, 2012), `https://www.owasp.org/index.php/`
    `Forced_browsing`
16. Bray, T.: Deep Linking in the World Wide Web (May 29, 2012),
    `http://www.w3.org/2001/tag/doc/deeplinking.html`
17. Jayaraman, K., Lewandowski, G., Talaga, P.G., Chapin, S.J.: Enforcing Request Integrity in Web Applications. In: Foresti, S., Jajodia, S. (eds.) Data and Applications Security and Privacy XXIV. LNCS, vol. 6166, pp. 225–240. Springer, Heidelberg (2010)

18. Balzarotti, D., Cova, M., Felmetsger, V., Vigna, G.: Multi-Module Vulnerability Analysis of Web-based Applications. In: CCS (2007)
19. Cova, M., Balzarotti, D., Felmetsger, V., Vigna, G.: Swaddler: An Approach for the Anomaly-Based Detection of State Violations in Web Applications. In: Kruegel, C., Lippmann, R., Clark, A. (eds.) RAID 2007. LNCS, vol. 4637, pp. 63–86. Springer, Heidelberg (2007)
20. Felmetsger, V., Cavedon, L., Kruegel, C., Vigna, G.: Toward Automated Detection of Logic Vulnerabilities in Web Applications. In: USENIX Security (2010)
21. Li, X., Xue, Y.: BLOCK: A Black-box Approach for Detection of State Violation Attacks Towards Web Applications. In: ACSAC (2011)
22. Vikram, K., Prateek, A., Livshits, B.: Ripley: Automatically Securing Web 2.0 Applications Through Replicated Execution. In: CCS (2009)
23. Bisht, P., Hinrichs, T., Skrupsky, N., Bobrowicz, R., Venkatakrishnan, V.N.: NoTamper: Automatic Blackbox Detection of Parameter Tampering Opportunities in Web Applications. In: CCS (2010)
24. Guha, A., Krishnamurthi, S., Jim, T.: Using Static Analysis for Ajax Intrusion Detection. In: WWW (2009)
25. Balduzzi, M., Gimenez, C.T., Balzarotti, D., Kirda, E.: Automated Discovery of Parameter Pollution Vulnerabilities in Web Applications. In: NDSS (2011)

Using Security Policies to Automate Placement of Network Intrusion Prevention

Nirupama Talele[1], Jason Teutsch[1], Trent Jaeger[1], and Robert F. Erbacher[2]

[1] Systems and Internet Infrastructure Security Lab
Pennsylvania State University
{nrt123,teutsch,tjaeger}@cse.psu.edu
[2] Network Security Branch
Army Research Laboratory
Robert.F.Erbacher@us.army.mil

Abstract. System administrators frequently use Intrusion Detection and Prevention Systems (IDPS) and host security mechanisms, such as firewalls and mandatory access control, to protect their hosts from remote adversaries. The usual techniques for placing network monitoring and intrusion prevention apparatuses in the network do not account for host flows and fail to defend against vulnerabilities resulting from minor modifications to host configurations. Therefore, despite widespread use of these methods, the task of security remains largely reactive. In this paper, we propose an approach to automate a minimal mediation placement for network and host flows. We use Intrusion Prevention System (IPS) as a replacement for certain host mediations. Due to the large number of flows at the host level, we summarize information flows at the composite network level, using a conservative estimate of the host mediation. Our summary technique reduces the number of relevant network nodes in our example network by 80% and improves mediation placement speed by 87.5%. In this way, we effectively and efficiently compute network-wide defense placement for comprehensive security enforcement.

1 Introduction

Many security administrators rely on network monitoring and Intrusion Detection and Prevention Systems (IDPS) to protect the hosts in their networks from remote adversaries. An Intrusion Detection System (IDS) inspect information flows to detect any malicious activity while Intrusion Prevention Systems (IPS) block remote access when such malicious activity is detected. When an IDS detects a malicious packet, either it may log the packet, allowing an administrator to take further action, or it may drop the packet to protect the receiving host process. Intrusion prevention systems can detect malware in packets and block denial of service attacks.

Despite widespread use of IDPSs and the deployment of host security mechanisms such as host firewalls and mandatory access control, the task of security practitioners is still reactive, responding to vulnerabilities as adversaries identify them. The IDPS can be classified as network based or host based, where

J. Jürjens, B. Livshits, and R. Scandariato (Eds.): ESSoS 2013, LNCS 7781, pp. 17–32, 2013.
© Springer-Verlag Berlin Heidelberg 2013

the former monitors the network for suspicious activity and the latter monitors a single host for malicious activity. We highlight two key reasons for the lack of security methods. First, network based monitoring is inherently incomplete, as only certain threats can be identified and/or blocked at the network without creating false positives. For example, only known malware is blocked by the network based IPS, so that no valid functionality is accidentally blocked. Second, systems do not coordinate network monitoring with host defenses, resulting in security loopholes. For example, the host may overlook remote threats that a network based IPS cannot block, or a compromised process may propagate a remote threat to another host.

In order to pro actively block remote adversaries, one must defend against all adversary accesses. We wish to monitor only a small number of mediation points so as to minimize resource costs. Researchers have previously explored methods to compute minimal cost placements for network based IPS configurations. These methods only focus on network flows [2,38,4] or utilize heuristic models of potential host vulnerabilities, as in methods for computing *attack graphs* [28,16,33], to guide placement choices. Security under these frameworks rely on heuristic models of possible attacks (e.g., host scans [26]), which may miss previously unseen attacks and remain vulnerable to minor host configuration changes. More specifically, we identify two major limitations in the current models of attacks: (1) they do not account for the hierarchical structure of network-connected resources into subnets, hosts, and individual processes to represent possible attack paths and (2) they fail to account for network defenses, such as labeled network connections [15,31,21].

Rather than just computing a minimal placement for network based IPS, we compute minimal mediation placements for the network and host flows. Such mediation must account for both the network based IPS as described above as well as the host mediation necessary to enforce a set of security properties. In this paper, we develop a method that utilizes the available security policies on commodity operating systems and networks to compute *mediation placements* which automatically block adversary access to security-critical data. Our new method, which places mediators based on authorized data flows and security policies at the network and host levels, yields robust defense. We find that those defenses that cannot be enforced by network based IPS must be implemented by host mediation. The proposed method produces this necessary host mediation, given the IPS capabilities and constraints.

We implement a two-stage algorithm for computing network-wide mediation placement. The idea behind this method is that, where possible, our method replaces host mediation with network based IPS and vice versa. In the first stage, we compute conservative host mediation for a worst-case set of remote threats for the hosts. Our mediation suffices to protect the hosts in any deployment without network defenses. In the second stage, we summarize data flows within each host by utilizing the conservative host mediation. We observed that host data flow summaries built from conservative host mediations reduce the number of nodes in the example network graph by over 80% and the number of edges by more than 60%. Using these summaries, the automated network-wide mediation placement time for the example network is reduced by 87.5% when the summaries are produced in advance, which is feasible in many cases. This result demonstrates the feasibility of automated, comprehensive, network-wide defense placement.

Contributions: The first contribution of this paper is combining the host data flow graphs as computed in [23] with the network data flow and generating a hierarchical encapsulated graph model representing information flows in both network and host in order to compute near minimal defense placement for the entire network. The second contribution is the optimization in the conservative host mediation placement, where the method replaces the host mediation with network based IPS wherever possible. And the third contribution of this paper is summarization of the data flow graph for each host utilizing the conservative host mediation in order to make our mediation placement technique feasible in large networks.

The remainder of the paper is organized as follows. Section 2 provides background on network defenses and host security mechanisms, and defines the mediation placement problem. Section 3 defines an information flow problem whose solution is also a solution for the mediation placement problem. Section 4 outlines the design of our method. Section 5 describes the evaluation platform and experimental results. Section 6 concludes the paper and identifies future work.

2   Background

In this section, we explain the need for mediation placement in networks of hosts and incorporate available security policies in the production of such placements.

2.1   Network Scenario

Figure 1 shows a typical modern networked application. Such applications consist of servers and their clients, where clients may be deployed in either wired or wireless networks. In many cases, clients and servers perform security-critical processing, assuming that the application data is protected from unauthorized modification or leakage and application data is available when necessary.

However, networked applications face a variety of threats. First, remote adversaries may launch attacks on processes that are accessible to the network at large. These processes often include custom programs such as PHP web applications on servers and unprivileged applications on clients with network access. Many system compromises now start by attacks on unprivileged processes. Second, processes on hosts within the network may launch attacks against other hosts. Such attacks may focus on system services by exploiting trust among hosts inside the network (and unauthenticated network protocols) or leverage the openness of wireless networks. Third, remote adversaries who are able to compromise an unprivileged process or trick users into installing untrusted data

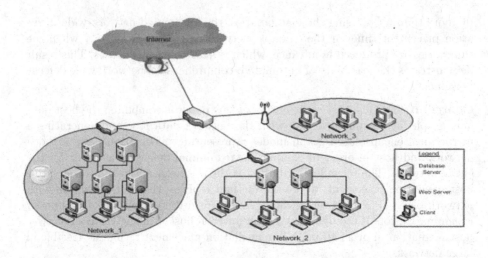

Fig. 1. Example Networked Application

may launch local exploits against more privileged processes to install root kits or obtain administrator privileges. Unfortunately, the number and variety of possible local exploits available to adversaries is beyond enumeration at present.

A security problem occurs when an adversary can execute a sequence of operations that results in access to unauthorized data or excessive use of data processing resources. Remote adversaries use access to available networks to find hosts with vulnerabilities necessary to obtain these goals. Modern systems block many trivial vulnerabilities, yet adversaries often find short sequences of compromises, including combinations of the unprivileged networked processes and local exploits described above, that lead to security breaches. As a result, security practitioners must block all *attack paths* [28,33,36], but the variety of possible attack paths (even short ones) has proven too complex for manual configuration.

2.2   Network Defense Placement Problem

A common method for protecting hosts from adversaries is IDPS. IDS can see all the network connections being utilized[1] and examine the transmitted network packets. Firewalls [5] now support powerful forms of deep-packet inspection to compare contents to attack signatures.

The common view asserts that proper defense placement depends on the type of network. In wired networks, IPS are often placed at network edges because all traffic must enter or leave via this choke-point. For wireless networks, IPS are placed on each node because other nodes in the wireless network may be untrusted. However, such placements may not effectively limit adversaries and/or

---

[1] Some uses of IPsec hide addressing information, but IDS is often placed where such information can be obtained, such as gateways.

may result in redundant monitoring. For wired networks, defenses at the edges ignores threats internal to the network, so if an adversary can compromise a single process on a single host they can then launch further attacks undetected. For wireless networks, per node monitoring may be unnecessary in some cases because certain kinds of attacks can be prevented at the edges. For example, a distributed denial of service attack can be thwarted external to the cell. In both types of networks, local exploits are invisible to the network based intrusion detection infrastructure, meaning the IDS placement has only a limited view of possible attack paths.

With the widespread deployment of mandatory access control (MAC) in commodity systems over the last ten years [32,30,40,37], adversary access within hosts can be restricted, although such restrictions do not block all adversary attack paths on the host. This MAC enforcement has been used primarily to confine network-facing daemons to prevent compromised root process from compromising the system at large. However, such enforcement does not prevent local exploits. Windows Mandatory Integrity Control [22] (MIC) is designed to prevent untrusted code (e.g., downloaded from the Internet) from modifying privileged resources, but does not prevent adversaries from tricking victims into reading untrusted data or upgrading untrusted code. As a result, we advocate development of a method to compute IPS placements that account for the host and network configurations.

Related Work: Many efforts have been made to verify the data flow in a policy [9,12,19,20], especially with the intention to assist in policy design and identification of unauthorized operation on network components. There are also many attack graph based methods to verify the network policies [13,33] which, as discussed above are heuristic based methods and may fail to represent the actual host behavior. However, the composed behavior of these arbitrary policies in hosts and networks is complex to analyze and the policies may interact in unpredictable ways. Researchers have recently explored methods to place minimal but comprehensive network defenses automatically by solving graph problems, such as vertex cover and multicuts [2,38,4,29]. While these problems are computationally complex in general, efficient greedy algorithms exist to produce effective solutions. Again, these methods make broad assumptions about the host that may misrepresent the actual attack paths within the host.

Project Goal: Our goal is to develop a method that computes comprehensive and minimal *mediation placement* for networks of hosts. Such method must utilize the host security policies in addition to those in the network to provide an accurate model of adversary access. Further, this method must account for the possible mediation capabilities in both network devices and on the hosts themselves. As discussed above, finding the minimal solution is difficult for known methods for general security policies, so the method must also leverage practical insights to produce effective approximate solutions.

3 Information Flow Problem

In this section, we show that the network and host mediation placement problem can be expressed as an information flow problem. Traditionally, an information flow problem is defined as follows:

Definition 1. *An* information flow problem, $\mathcal{I} = (\mathcal{G}, \mathcal{L}, \mathcal{M})$, *consists of the following concepts:*

1. *A directed data flow graph* $G = (V, E)$ *consisting of a set of nodes* V *connected by edges* E.
2. *A lattice* $\mathcal{L} = \{L, \preceq\}$. *For any two levels* $l_i, l_j \in L$, $l_i \preceq l_j$ *means that* l_i '*can flow to*' l_j.
3. *A level mapping function* $M : V \to \mathcal{P}^L$ *where* $\mathcal{P}^L$ *is the power set of* L *(i.e., each node is mapped either to a set of levels in* L *or to* $\emptyset$*)*.
4. *The lattice imposes security constraints on the information flows enabled by the data flow graph. Each pair* $u, v \in V$ *s.t.* $[u \hookrightarrow_G v \land (\exists l_u \in M(u), l_v \in M(v).\ l_u \npreceq_{\mathcal{L}} l_v)]$, *where* $\hookrightarrow_G$ *means there is a path from* u *to* v *in* G, *represents an* information flow error.

It has been shown that information flow errors in programs [25] and MAC policies [14,35,3] can be automatically found using such a model.

However, resolving such information flow errors has been a complex manual task. In general, information flow errors can be resolved by changing the data flow graph (e.g., removing nodes and/or edges) or adding *mediation* to change the level of data propagated by information flows. However, changing the data flow graph is difficult in practice because it implies a change in the operations a system may perform, which may prevent one or more components from functioning correctly.

Researchers have developed methods to generate minimal mediation to automatically resolve information flow errors by independently proposing graph-cut-based solutions for MAC policies of programs [17,18] and systems [34]. A graph-cut solution identifies a minimum cost set of mediators R defined as $R = \{((u, v), l) \mid (u, v) \in E \land l \in L\}$, where edge (u, v) is a *cut-edge* and l is the data *security level* sent by u to v due to mediation, resulting in the mapping $M_v = v \to l$ being assigned to the edge's destination v. That is, each cut-edge relabels the information received by v from u to l. This set of cut-edges in R resolves all information flow errors in the information flow problem $\mathcal{I}$, according to the Cut-Mediation Equivalence [18].

In a recent paper [23], we extended the basic graph cut problem to account for the integration of independent components into a composite and coherent data flow graph. The method also accounted for the constrained ability of partially trusted components to mediate information flows, and proposed a strategy for placing mediation that implemented a classical integrity model [6]. In general, the information flow policy may be a partially ordered set of permissions, and therefore an optimal mediation solution requires us to solve a multicut problem.

In light of the intractability of multicut [7], we apply a greedy method to produce an approximate solution.

We find that by solving this information flow problem, we can also produce a network defense placement. Further, by accounting for host and network data flows comprehensively, we produce a network defense placement that accounts for attack paths more accurately. However, the unwieldy size of the combined host and network data flows make obtaining these solutions computationally infeasible. Indeed, the data flow graph for each host consists of 2000–3000 nodes and 6000–12000 edges. Given that our greedy multicut graph cut algorithm runs in $O(|L| \cdot n^3)$ where n is the number of nodes in the data flow graph, only networks with a modest number of hosts can be considered in practice. In this paper, we develop an approximate (greedy) solution along with host summaries for network defense placement that accurately accounts for network and host data flows.

Assumptions. The key assumption in this work is that the devices, operating systems, and programs that enforce security policies, do so correctly. This is a significant assumption given the size and complexity of such components, but it is also a standard assumption in modern computing systems. Specifically, we assume that the devices, operating systems, and programs that enforce security policies satisfy the *reference monitor concept* [1], which requires that a reference validation mechanism (i.e., MAC enforcement) "must always be invoked" upon a security-sensitive operation, "must be tamper proof," and must be "small enough to be subject to analysis and tests, the completeness of which can be assured," which implies correctness. The reference monitor concept is certainly the goal of several commodity reference validation mechanisms, although they do not meet the latter of these requirements.

4   Design

In this section, we design a method for computing mediation placements for network and host data flows. First, we compute the conservative mediator placement required for each unique host configuration in the network. Second, we produce summaries of the resultant host information flows, accounting for the placed mediation. Third, these summaries are used to produce a more feasible information flow problem

4.1   Host Mediation

In the first step, we describe a method to compute a conservative host mediation that would protect the host when there is no network defense. Our fundamental task is to constrain the information flow problem to the union of sufficient subproblems, as we use a graph-cut method described in [24] to compute these sub mediations.

A host is defined by its internal data flows and network connections. Many commodity systems are now deployed with *mandatory access control* (MAC)

policies [32,30,37,40]. MAC policies define the legal operations of subjects on objects in the host. We compute the internal data flow graph among subjects and objects of a host from its MAC policy using well-known techniques [39,14,35,3].

In addition to the host data flows, some subjects in the host may have access to the network. The combination of the MAC policy data flows and network data flows for host subjects forms the host data flow graph $G = (V, E)$. The network access is represented by sets of input and output nodes, $I \subseteq V$ and $O \subseteq V$ respectively, and edges that identify which MAC policy subject nodes can access the network nodes. For the computation of individual host mediation, each element in I has an indegree of 0, and each element of O has an outdegree of 0. For many firewall rules, the port uniquely identifies the MAC policy subject that can access the network, but for some client ports such connections may be ambiguous. These must be identified before analysis.

To produce an information flow problem, we must produce a lattice policy and map the lattice levels to the appropriate nodes in the data flow graph using a mapping function, see Definition 1. For OS distributions which specialize in single applications (e.g., web server, database, etc.), we associate lattice levels with the kernel and application labels in the MAC policy by specifying such mapping functions [23]. The input nodes are mapped to the lattice level for the expected input data. Typically, we assign the input nodes to the level for remote adversaries because most network inputs are untrusted. The input node mapping must represent the worst-case scenario for the host, as there is no network based defenses at this stage.

In Section 3, we stated that a solution consists of a set of mediator edges $R = \{((u, v), l) \text{ where } (u, v) \in E \text{ and } l \in L\}$ and finding such a solution is a multicut problem [7] for a general lattice. Finding a minimal solution to the Information Flow problem is NP-hard. For this reason, we employ a greedy algorithm rather than attempting to obtain an exactly minimal solution to the information flow problem. The greedy method solves a min-cut problem [10] for each prohibited pair of lattice levels and outputs the union of these cuts. If we have k such pairs, then the greedy solution obtained is no greater than k times the optimal cut. The reason for this is that the optimal cut can be no smaller than the size of the minimal cut for an individual pair, and the algorithm makes k such cuts. We solve the sub problems by topologically sorting the lattice to take advantage by reuse of solutions, as described in [23]. By solving the corresponding information flow problem, we obtain a host mediation R at the program entry points (i.e., program instructions that invoke the system call library [13]) necessary to protect host processes from the specified remote threats. By construction, our greedy algorithm contains, for each prohibited lattice level pair, a set of edges whose deletion separates the corresponding nodes in the network graph.

4.2   Host Summaries

In this section, we use the host mediation placement to summarize the flows within the hosts to reduce the size of the network-wide information flow problem.

A summary of a host data flow graph consists of the data flows from the host's input nodes to its output nodes and shows how data received by this host is propagated to other hosts via its output. We define the function $Reach(S, T) = \{(s, t) \mid s \in S, t \in T, s \neq t \wedge (s \hookrightarrow_G t)\}$. In general, the data flows through a directed graph G can be summarized as $G' = (V', E')$, where $V' = I \cup O$ and $E' = Reach(I, O)$.

To accurately capture information flows through a host, we must also account for the mapping function M, which defines where data of particular security levels host imports, and the host mediator placement R computed in the previous section, which defines where the security level of the data on a particular data flow is changed. Since the mapping function and mediators affect the security level of the data the host produces, our summary must take these into account[2].

First, we leverage the knowledge that the mapping function $M : V \to \mathcal{P}^L$ identifies a set of nodes $A \subseteq V$ that are mapped to lattice levels in L. Secondly, application of a mediator also changes the information flows through the host. A mediator $((x, y), l) \in R$ does two things: (1) it filters the flow through edge $(x, y) \in E$ in the graph and (2) it maps a new level $l \in L$ to the node y. That is, a mediator causes the receiving node of an edge to receive data mapping to the lattice level of the mediator. As a result, we retain the mediator edges (R_x, R_y) in the summary. This results in the following definition of a summary graph.

Definition 2. *For graph G given the input sets I, O, R_x, R_y and A such that R_x and R_y are sets of source and target nodes of mediator edges in R respectively. A summarized data flow graph G' is a directed graph $G' = (V', E')$, where $V' = I \cup O \cup A \cup R_x \cup R_y$ and $E' = Reach(I, O) \cup Reach(R_x, R_y) \cup Reach(I, R_x) \cup Reach(R_y, A) \cup Reach(A, O)$.*

That is, the outputs are either based on the input data (edges in $Reach(I, O)$) or based on the mapped data (edges in $Reach(A, O)$) which may be combined with some input data and mediators (edges in $Reach(I, R_x)$ and $Reach(R_y, A)$). Note that if there is a flow from node $i \in I$ and a flow from node $a \in A$ that merge at some node $x \notin O$, the correct output flows will still be produced. Either there is a path from x to node $o \in O$ causing $Reach(I, O) \cup Reach(A, O)$ to include edges (i, o) and (a, o) replicating the merge, or neither reaches a node in O meaning that no edges are produced. Figure 2 represents the summarization of a host graph where all the relevant paths from inputs to mediators and mapped nodes, and from mapped nodes to outputs are retained.

The claim is that the summarized data flow graph includes all the edges necessary to compute the output information flow of the host accurately.

Theorem 3. *The summarized data flow graph G' constructed using Definition 2 for a directed graph G with a particular mapping function M, lattice $\mathcal{L}$, and set of mediators R produces the same information flows to all nodes in O as G, regardless of the security levels mapped to the inputs nodes I.*

---

[2] When data of different security levels is combined, the resultant security level is the least-upper bound of the input levels, as defined by Denning's Lattice Model [8].

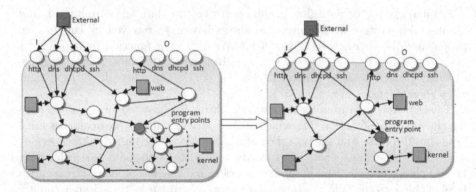

Fig. 2. Graph Summarization: Where the box represents the security level mapping to the node, green is trusted and red is untrusted. The blue nodes are nodes which can be selected for mediation(R_x).

Proof. The information flows that reach nodes in O in G are a combination of flows propagated from the inputs to the outputs ($Reach(I, O)$), flows from the mapped nodes and mediators to the outputs ($Reach(A, O)$), flows from inputs to the mediator nodes ($Reach(I, R_x)$), and flows from mediator nodes to other mapped nodes ($Reach(R_y, A)$). As long as we include all the relevant flows that alter the security level while computing the summary graph G', the flows $Reach(I, O)$ and $Reach(A, O)$ will capture only the flows to the nodes in O in summary graph G'.

4.3 Network-Wide Mediation Placement

Given the summarized data flow graphs for each host and the network policies that define the data flows among hosts, it is possible to compute a IPS placement. The goal is to replace the host mediation with network IPS wherever possible such that this replacement reduces the overall number of mediators required. To do so, we compute a cut solution for an information flow problem built from the composition of network and summarized host data flows. The composite result is guaranteed to require no more than the number of mediators in the conservative host mediation for all network hosts.

We solve the IPS placement problem by building a second information flow problem covering the network flows and summarized hosts (see Definition 1). The data flow graph of this information flow problem is now a combination of summarized host data flow graphs (from the previous section) and the network data flows that connect hosts. The network data flows are derived from the possible network connections in the particular network type (e.g., wired or cellular) and the firewall policies of the network (e.g., edge servers) and hosts.

The recursive method for composing the hierarchical model for network and host data flows is presented in Algorithm 1. The algorithm performs a postorder precessing where the parent node can only be processed after the child nodes in order to add the required edges. As networks of hosts are organized hierarchically and flows between hosts or networks are encapsulated, we use an encapsulated, hierarchical graph model to represent the composite data flow graph [27,11]. An advantage of such a graph model is that we can plug summaries of hosts and even networks into a data flow graph easily. The input and output nodes of child host are projected at the parent node to provide an interface with the external world. The child elements are connected as per the network configuration provided. For instance all hosts are interconnected in the wireless network $Network\_3$, while in the wired network $Network\_1$ all child hosts are not necessarily interconnected and communicate as defined by the topology.

Algorithm 1. Generate Hierarchical Network Graph

Input: *host* contains all the allowed policy flows in a host and *net_conf* contains the allowed flows by firewall and network configuration files
Output: Hierarchical network graph model $Network\_M$
1: **function** $Gen\_Network\_Graph(host, net\_conf)$
2:   //Recursively build all child hosts
3:     $N = host.Children.Count;$
4:     **for** $i = 0$ *to* N **do**
5:         $Network\_M.child[i] = Gen\_Network\_Graph(host.child[i]);$
6:     **end for**
7:   //Process parent node: Generate Network Edges
8:     $Add\ interface\ for\ child\ I/O\ ports$
9:     **for** $(all\ u, v\ in\ Network\_M.V)$ **do**
10:         **if** $((u, v) \in host.flows$ and $(u, v) \in net\_conf.flows$ **then**
11:             $Network\_M.E = Network\_M.E \cup (u, v);$
12:         **end if**
13:     **end for**
14:     $Network\_M.Graph = M(Network\_M,\ L);$ //lattice mapping function.
15:     $Summarize(Network\_M);$ //Summarize host graph
16: **end function**

We configure the lattice and mapping function for the information flow problem as follows. Since network devices are not really visible to the hosts, they do not introduce any new lattice levels or mappings[3]. However, we do need to know the sources of network adversaries in order to map the threats (i.e., adversarial security levels) to their actual network locations.

We note that not all network devices may be capable of mediating all host requirements. For example, a network IPS may scan for known malware, but a host process that is accessible to an adversary must protect itself from any malicious input, known or unknown. To express such limitations in solving information flow problems, we forbid certain edges from being mediators [23]. Such constraints associate an edge with a lattice level and prevent any mediator assigned to that edge from declassifying data above that lattice level.

---

[3] Of course, we may want to place mediation to protect the network devices, but that is not the focus in this paper.

As a result, the set of network devices may not be capable of realizing a given solution to the information flow problem. Thus, we retain the possibility of using the host mediators in addition to network IPS to solve the information flow problem. These two sets of locations are the only possible mediation locations for this information flow problem. In the worst-case, the conservative host mediation produced in the previous section will be used to protect the system, but we apply network IPS where it reduces the cost of host mediation.

To compute a network-wide mediator placement, we solve the information flow problem above by computing a set of mediators that resolve all information flow errors for the network-wide data flow graph. Using the summarized data flow graphs and the conservative host mediations computed in Section 4.1, we see that the method shifts host mediation to network IPS where possible. $\mathcal{R}$ is the union of all the conservative host mediations for all the systems hosts in the network. Since we computed $\mathcal{R}$ using the worst-case input mapping, $\mathcal{R}$ is a solution to the network-wide information flow problem. The claim is that any network IPS that leverages the above solution to this information flow problem only reduces the amount of overall mediation required.

Theorem 4. *Given a conservative host mediation for a set of hosts $\mathcal{R}$ consisting of T edges, a mediation placement can be found that solves the information flow problem containing these hosts in a network data flow graph constructed as described above and the number of mediators in that solution is less than $|T|$.*

Proof. A cut problem is created for each lattice level $l_i \in L$ mediated by the conservative host mediations $\mathcal{R}$, resulting in $|L_\mathcal{R}|$ cut problems for $L_\mathcal{R} \subseteq L$ levels mediated total. A conservative host mediation has a set of edges T_i that mediate to level l_i. Note that this set forms a valid cut solution of size $|T_i|$ for the cut problem to l_i. Thus, the union of these cut solutions is a valid multicut that resolve all information flow errors of size $\sum_{i=1}^{|L_\mathcal{R}|} |T_i|$. However, any solution of edges S_i for lattice level l_i must be of size $|S_i| \leq |T_i|$, because the cut problem solution is exact. Thus, the sum of these sizes of the cut solutions for each level l_i, $|S_i|$, must be no greater than $|T|$.

5   Evaluation

In this section, we describe the prototype implementation for a sample networked application. We base the analysis of individual host mediation on our previous work [23] with a new hierarchical modeling and host summarization module implemented in C++. Figure 1 shows the experimental setup with 16 hosts in a sample network configuration. The network hosts include a collection of web servers, database servers, and web clients. Each of the network host is a VM that runs the Linux 2.6.31-23-generic kernel and enforce SELinux refpolicy 2.20120725 [32] with different module configurations. Each host policy is different and supports distinct set applications. The web server VM runs an Apache web server and web application, database VM runs MySQL and, client VMs run a web browser. Network 1 and 2 are both wired networks with one at a higher security

Table 1. Impact of summarization on individual host graphs

Host type	Nodes (G)	Edges (G)	Nodes (G')	Edges (G')	Reduction Nodes	Reduction Edges	Time (sec)
Web Server(5)	2050	6660	309	2509	85%	62.3%	192.1
Database Server(2)	2578	10071	359	2179	86%	78.3%	267.26
Web Client(8)	2978	11499	479	2332	84%	79.7%	302.57

Table 2. Performance gain with summarized hosts for sample networked application

Network	Nodes	Edges	Mediators	Time(min)
Before Summarization	39,235	145,476	4,745	34.19
After Summarization	6,176	36,031	4,745	4.27
Reduction Percentage	84.2%	75.2%	N/A	87.5%

level than the other. The hosts in Network 2 are not directly connected to each other and communicate via the network node, while some hosts in network 1 can directly talk to each other. Network 3 is a wireless network in which all nodes may communicate with each other as one would expect in a situation where signals are broadcast into a public space.

We use individual firewall (iptables) rules to determine a host's interaction with the outer world and the network configuration files to determine the overall organization of hosts in this networked application. We now describe the impact of summarization on the computation of network-wide mediation placement considering the optimizations from the conservative host mediation. We also demonstrate the joint optimized analysis of host mediations and network IPS.

We compute the summary graph for our hosts as described in Section 4.2. Table 1 shows the reduction in the graph for individual hosts on computing the host summary with conservative mediation placement. For each of the hosts, the number of nodes are reduced by over 80% and the number of edges are reduced by over 60%. We incorporate the conservative mediation placement, since only about 10% of the nodes can be removed without it, as nearly all processes may be capable of performing some kind of mediation. Table 1 specifies the average time needed for generating the summary graph for each kind of host. The summarization needs to be computed only once after which the summarized host can be reused multiple times for analyzing in different network environments.

Table 2 shows the impact of summarization on computing time for mediation placement. We compute a mediation placement before and after host summarization and find an 87.5% reduction in the placement analysis time for the example network. The total number of mediation placements is 4745, out of which only 613 mediators are unique. A unique mediator is counted only once even if it is used in different hosts. A program may appear in multiple hosts, thus filtering such repetition gives us a reduction of 87% in terms of the effort required for deploying the mediators.

Table 3 shows the average reduction in the host mediation for each type of host when network IPS is available. It also shows average time required to generate the conservation mediations for each type of host. In this example, we aim to compute the necessary mediation to block denial of service attacks as an information flow problem. The conservative host mediation shows the

Table 3. Mediator placement: conservative vs in a network

Host type	Mediators (Conservative)	Mediators (Network)	Time-Conservative mediators (sec)
Web Server(5)	258	214	62.86
Database Server(2)	308	229	80.55
Web Client(8)	429	283	93.55

number of program entry points necessary to block paths from the network inputs to protect the application (e.g., web server, database, web clients) and system resources (e.g., critical kernel files). We assume that the network IPS can block application-specific malware directed at the application over the network. As only a small number of paths exists directly from the network to these applications, and because malware may compromise system processes by passing through applications, we can block only a fraction of the threats with network IPS, even with complete malware detection at the IDS. This result can help system administrator identify what defenses are needed in the host and what can be entrusted to network in a particular configuration. The experiment can be performed on various network states at static time and it is part of future work to adapt the analysis to handle dynamic network. The method determines the placement, such that no untrusted data can reach a trusted object without going through the appropriate mediator, enforcement of which has to satisfy the reference monitor concept as expressed in Section 3.

6   Conclusion

In this paper, we proposed a method for computing a minimal placement for network defenses among network and host flows. While networks with multiple hosts can have many flows, we demonstrated a feasible approach which views network IPS as a replacement for certain host mediations. We designed an algorithm based on summarization of host flows and a conservative estimate of the required host mediation that reduced the size of the information flow problem by more than 80% and mediation placement computation time by 87.5% in our prototype network. Thus, our method provides automated network-wide defense placements which comprehensively enforce security. We also observe that hosts performing similar functionality with similar security requirements use mediations at the same entry points. Our experimental results indicate that the redundancy in networks of systems offer future opportunities for host based summarization.

References

1. Anderson, J.P.: Computer security technology planning study, vol. II. Technical Report ESD-TR-73-51, Deputy for Command and Management Systems, HQ Electronics Systems Division (AFSC) (October 1972)
2. Breitbart, Y., Dragan, F., Gobjuka, H.: Effective monitor placement in internet networks. Journal of Networks (2009)

3. Chen, H., Li, N., Mao, Z.: Analyzing and comparing the protection quality of security enhanced operating systems. In: NDSS (2009)
4. Chen, X., Kim, Y.-A., Wang, B., Wei, W., Shi, Z.J., Song, Y.: Fault-tolerant monitor placement for out-of-band wireless sensor network monitoring. Ad Hoc Networks 10(1), 62–74 (2012)
5. Cheswick, W.R., Bellovin, S.M., Rubin, A.D.: Firewalls and Internet Security; Repelling the Wily Hacker, 2nd edn. Addison-Wesley, Reading (2003)
6. Clark, D.D., Wilson, D.: A comparison of military and commercial security policies. In: IEEE Symposium on Security and Privacy (1987)
7. Dahlhaus, E., Johnson, D.S., Papadimitriou, C.H., Seymour, P.D., Yannakakis, M.: The complexity of multiterminal cuts. SIAM J. Comput. 23, 864–894 (1994)
8. Denning, D.: A lattice model of secure information flow. Communications of the ACM 19(5), 236–242 (1976)
9. Dragoni, N., Massacci, F., Naliuka, K., Siahaan, I.: Security-by-Contract: Toward a Semantics for Digital Signatures on Mobile Code. In: López, J., Samarati, P., Ferrer, J.L. (eds.) EuroPKI 2007. LNCS, vol. 4582, pp. 297–312. Springer, Heidelberg (2007)
10. Ford, L.R., Fulkerson, D.R.: Flows in Networks. Princeton University Press (1962)
11. Fritz, D.G., Sargent, R.G.: An overview of hierarchical control flow graph models. In: Proceedings of the 27th Conference on Winter Simulation, WSC 1995, pp. 1347–1355. IEEE Computer Society, Washington, DC (1995)
12. Hicks, B., Rueda, S., St. Clair, L., Jaeger, T., McDaniel, P.: A logical specification and analysis for SELinux MLS policy. ACM Transaction on Information and System Security 13(3) (2010)
13. Howard, M., Pincus, J., Wing, J.: Measuring relative attack surfaces. In: Proceedings of Workshop on Advanced Developments in Software and Systems Security (2003)
14. Jaeger, T., Sailer, R., Zhang, X.: Analyzing integrity protection in the SELinux example policy. In: USENIX Security Symposium (August 2003)
15. Jaeger, T., Butler, K., King, D.H., Hallyn, S., Latten, J., Zhang, X.: Leveraging IPsec for mandatory access control across systems. In: Proc. 2nd Intl. Conf. on Security and Privacy in Communication Networks (August 2006)
16. Jha, S., Sheyner, O., Wing, J.: Two formal analyses of attack graphs. In: Proceedings of the 15th IEEE Workshop on Computer Security Foundations, pp. 49–63. IEEE Computer Society, Washington, DC (2002)
17. King, D., Jha, S., Jaeger, T., Jha, S., Seshia, S.A.: Towards automated security mediation placement. Technical Report NAS-TR-0100-2008, Network and Security Research Center, Department of Computer Science and Engineering, Pennsylvania State University, University Park, PA, USA (November 2008)
18. King, D., Jha, S., Muthukumaran, D., Jaeger, T., Jha, S., Seshia, S.A.: Automating Security Mediation Placement. In: Gordon, A.D. (ed.) ESOP 2010. LNCS, vol. 6012, pp. 327–344. Springer, Heidelberg (2010)
19. Massacci, F., Siahaan, I.: Matching Midlet's security claims with a platform security policy using automata modulo theory. In: Proceedings of NordSec (2007)
20. McDaniel, P., Prakash, A.: Methods and limitations of security policy reconciliation. ACM Trans. Inf. Syst. Secur. (2006)
21. Morris, J.: New Secmark-based network controls for SELinux, http://james-morris.livejournal.com/11010.html
22. MSDN. Mandatory Integrity Control (Windows), http://msdn.microsoft.com/

23. Muthukumaran, D., Rueda, S., Talele, N., Vijayakumar, H., Jaeger, T., Teutsch, J., Edwards, N.: Transforming commodity security policies to enforce Clark-Wilson integrity. In: ACSAC (2012)
24. Muthukumaran, D., Jaeger, T., Ganapathy, V.: Leveraging "choice" to automate authorization hook placement. In: CCS 2012: Proceedings of the 19th ACM Conference on Computer and Communications Security. ACM Press, Raleigh (2012)
25. Myers, A.C., Liskov, B.: A decentralized model for information flow control. ACM Operating Systems Review 31(5), 129–142 (1997)
26. Nessus Vulnerability Scanner, http://www.nessus.org/
27. Noble, J., Biddle, R., Tempero, E., Potanin, A., Clarke, D.: Towards a model of encapsulation. Presented at the ECOOP 2003 IWACO Workshop on Aliasing, Confinement, and Ownership (publications) (2003), http://www.mcs.vuw.ac.nz/comp
28. Noel, S., Jajodia, S.: Advanced vulnerability analysis and intrusion detection through predictive attack graphs. In: Critical Issues in C4I, Armed Forces Communications and Electronics Association (AFCEA) Solutions Series. International Journal of Command and Control (2009)
29. Noel, S., Jajodia, S., O'Berry, B., Jacobs, M.: Efficient minimum-cost network hardening via exploit dependency graphs. In: ACSAC (2003)
30. Novell. AppArmor Linux Application Security, https://www.suse.com/support/security/apparmor/
31. NetLabel - Explicit labeled networking for Linux, http://www.nsa.gov/research/selinux/
32. Security-enhanced linux, http://www.nsa.gov/research/selinux/
33. Ou, X., Boyer, W.F., McQueen, M.A.: A scalable approach to attack graph generation. In: CCS (2006)
34. Pike, L.: Post-hoc separation policy analysis with graph algorithms. In: Workshop on Foundations of Computer Security (FCS 2009). Affiliated with Logic in Computer Science (LICS) (August 2009)
35. Sarna-Starosta, B., Stoller, S.D.: Policy analysis for Security-Enhanced Linux. In: WITS (April 2004)
36. Sheyner, O., Haines, J.W., Jha, S., Lippmann, R., Wing, J.M.: Automated generation and analysis of attack graphs. In: IEEE Symposium on Security and Privacy, pp. 273–284 (2002)
37. Sun Microsystems. Trusted Solaris operating environment - a technical overview, http://www.sun.com
38. Tang, Y., Daniels, T.E.: On the economic placement of monitors in router level network topologies. In: The Workshop on the Economics of Securing the Information Infrastructure (2006)
39. Tresys. SETools - Policy analysis tools for SELinux, http://oss.tresys.com/projects/setools
40. Watson, R.N.M.: TrustedBSD: Adding trusted operating system features to FreeBSD. In: Proceedings of the FREENIX Track: 2001 USENIX Annual Technical Conference, pp. 15–28 (2001)

Idea: Callee-Site Rewriting
of Sealed System Libraries

Philipp von Styp-Rekowsky[1], Sebastian Gerling[1], Michael Backes[1,2],
and Christian Hammer[1]

[1] Saarland University, Saarbrücken, Germany
[2] Max Planck Institute for Software Systems (MPI-SWS), Saarbrücken, Germany
lastname@cispa.uni-saarland.de

Abstract. Inline reference monitoring instruments programs in order
to enforce a security policy at runtime. This technique has become an
essential tool to mitigate inherent security shortcomings of mobile platforms like Android. Unfortunately, rewriting all calls to security-relevant
methods requires significant space and time, in particular if this process
is performed on the phone. This work proposes a novel approach to inline
reference monitoring that abstains from caller-site instrumentation even
in the case where the monitored method is part of a sealed library. To
that end we divert the control flow towards the security monitor by modifying references to security-relevant methods in the Dalvik Virtual Machine's internal bytecode representation. This method is similar in spirit
to modifying function pointers and effectively allows callee-site rewriting.
Our initial empirical evaluation demonstrates that this approach incurs
minimal runtime overhead.

Keywords: Android, inline reference monitoring, sealed libraries.

1 Introduction

Mobile devices nowadays store a plethora of sensitive information about our
private and business life. Often, this information can be accessed in predefined
locations like an address book, photo folders, etc., where an attacker can easily locate them. However, this information is often not properly protected. For
example, on Android users have no choice but to grant an app all requested
permissions at install time, and these permissions cannot be revoked later on.
At the same time, these permissions are coarse-grained and hard to understand
for the average user. In the past, several incidents have been reported, where private information was deliberately leaked to the servers of an app. Even widely
used apps like Facebook, LinkedIn, and WhatsApp used to clandestinely send
the phone's whole address book to their servers to mine for possible contacts.

Inline reference monitoring [5] enforces a security policy at runtime by rewriting the original program to check the policy before executing security-critical
operations. In contrast, external reference monitors [11,2,6] require changes to
the firmware of a mobile phone and thus have limited practicability. Recently,

J. Jürjens, B. Livshits, and R. Scandariato (Eds.): ESSoS 2013, LNCS 7781, pp. 33–41, 2013.
© Springer-Verlag Berlin Heidelberg 2013

several variants of inline reference monitoring [1,9,3] have been proposed to mitigate the shortcomings of Android's security system. While some enforce more general security policies, all of them allow dynamic permission revocation in order to regain control over the sensitive data accessible to apps. Untrusted apps are rewritten to invoke a security monitor before each security-sensitive operation, which is typically a call to a method defined in Android's system libraries. The monitor checks whether the (current) security policy allows the attempted operation: In the positive case it lets the original call proceed, while a negative decision blocks the security-sensitive operation. In the latter case, it returns a mock value to prevent the app's termination due to an exception, if necessary. This variant of inline reference monitoring is called *caller-site rewriting*, as any call to a security-sensitive operation must be instrumented.

An alternative reference monitor style, called *callee-site rewriting* is far less invasive, as it only instruments the entry of the security-sensitive method itself instead of all the invocation points. On top of that, it also monitors calls that are not statically determinable, such as reflective calls or calls from the Java Native Interface (JNI). Unfortunately, callee-site rewriting is not feasible for almost all security-relevant code, which is often defined in sealed libraries (i.e. which cannot be modified) and loaded before any client code executes. Thus static rewriting of these libraries is impossible.

Our contribution is that we enable callee-site rewriting for sealed libraries. We achieve this by diverting control flow in the virtual machine. This insight is based on the observation that the VM-internal data structures that represent the libraries in memory are modifiable. Therefore, it is possible to alter the control flow by modifying the reference to the library method's bytecode, which reroutes a call to this method to another piece of bytecode. For the purpose of inline reference monitoring, we relay an invocation of a security-relevant method to a method that checks whether the security policy allows the original invocation. If this is *not* the case, we simply return (a mock value); otherwise we invoke the original method with the original parameters. As we are altering references to Java bytecode, we have access to all the parameters of the original method call when checking the security policy, just like the caller-site instrumentation does. At the same time, if an app invokes the security-relevant method from JNI or via reflection, we will still monitor this call as it jumps into the monitor, as well. We are not aware of any previous work that modified references to bytecode in a virtual machine in order to divert control flow to a different functionality.

In more detail, we make the following contributions:

1. We propose to rewrite references to the bytecode of security-relevant methods in-memory in order to achieve the same effects as callee-site rewriting does. To that end, we invoke a native method at program entry that diverts control flow from security-sensitive methods to our monitor methods. We achieve this by modifying the reference to the bytecode of the security-sensitive method and storing the original reference inside the monitor, which effectively makes the original method available to our monitor only.

2. Our in-memory callee-site rewriting only requires minimal instrumentation of an app that needs to be protected by a security policy. In practice, all entry points to the program need to ensure that the references have already been altered. Otherwise, a piece of native code is invoked that modifies these references.
3. In-memory rerouting of security-sensitive methods allows dynamic policy updates and is more efficient than static program rewriting, as it only alters the references of methods that the policy currently protects. Static rewriting would either need to instrument all potentially security-relevant methods or to re-instrument whenever the policy is modified. On top of that, our technique is less invasive, which facilitates on-the-phone instrumentation and minimizes possible conflicts with the original application.
4. We demonstrate the feasibility of our proposed technique by a prototypical implementation. Initial micro-benchmarks show that the dynamic overhead of this technique is minimal and negligible in a practical application.

2 Background

Runtime policy enforcement for third-party applications cannot be easily integrated into unmodified Android systems. Android's security concept strictly isolates different applications installed on the same device to prevent apps from interfering with each other at runtime. Furthermore, applications cannot gain elevated privileges to observe the behavior of other applications. Communication between apps is only possible via Android's inter-process communication (IPC) mechanism. However, such communication requires both parties to cooperate, rendering this channel unsuitable for a generic runtime monitor.

Recently, several approaches tackled this problem by following an approach pioneered by Erlingsson and Schneider [5] called *inline reference monitor* (IRM). The basic idea is to rewrite an untrusted application such that the code that monitors the application is directly embedded into its code. To this end, IRM systems incorporate a *rewriter* or *inliner* component that injects additional security checks, called guards, at critical points into the application bytecode. A guard can be injected into the control flow at different positions, but clearly, such a guard should be executed before the critical functionality is executed. There are two semantically equivalent approaches to IRM, as presented in Figure 1: *Caller-site-rewriting* (a) adds the guard before every critical call, while *callee-site rewriting* (b) injects the guard into the entry of the critical function itself. The latter is usually more efficient, since the guard only needs to be injected once. Unfortunately, Android's system libraries are sealed so that inlining the guards into a library is impossible. In order to achieve the same effect as traditional callee-site rewriting we divert all function calls from the security-critical library method to our security guard (see Figure 1(c)). Once the guard allows the execution, the original library function is invoked. This redirection, however, incurs an additional method call.

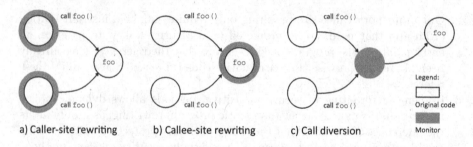

a) Caller-site rewriting   b) Callee-site rewriting   c) Call diversion

Fig. 1. Rewriting approaches

The injected security guards can now efficiently enforce a *security policy*. To actually enforce a policy, the monitor may suppress or alter calls to security-relevant functionality, or even terminate the program if necessary.

In the IRM context, a policy is typically specified by means of a security automaton that defines which sequences of security-relevant events are acceptable. Such policies have been shown to express exactly the policies enforceable by run-time monitoring [12]. Ligatti et al. differentiate security automata by their ability to enforce policies as they manipulate the trace of the program [10]. Some IRM systems [5,4] implement *truncation automata*, which can only terminate the program if it deviates from the policy. However, this is often undesirable in practice. Ligatti et al. [10] formulate the notion of *edit automata*, which can transform the program trace by inserting or suppressing events. Monitors based on edit automata are able to react gracefully to policy violations, e.g. by suppressing an undesired method call and returning a mock value, while allowing the program execution to continue.

On top of providing an elegant security policy enforcement mechanism, a key aspect of IRM-based security solutions is ease of deployment. User's need to be able to install the security system without requiring expert knowledge (e.g. gaining root access or changing the smartphone firmware). Our prior work AppGuard [1] demonstrates that rewriting apps directly on the phone and subsequently installing the instrumented apps is possible without modifying the base operating system or requiring root access.

3 Implementation

Our approach is based on diverting function calls to system libraries to functions in our own library that first perform a security check. The diversion is achieved by replacing the reference to a method's bytecode in the VM's internal representation (e.g. a virtual method table) with the reference to our security guard. Our security guards reside in an external library which is dynamically loaded on application startup. Therefore, we do not need to reinstrument the app when the security policy is modified. Additionally, we store the original reference in order to access the original function later on, e.g., in case the security check grants the permission to execute the security-critical method. In order

```
public class Main {
    public static void main(String[] args) {
        A.foo(); // calls A.foo()

        MethodHandle A_foo = Instrumentation.replaceMethod(
            "Lcom/test/A;->foo()", "Lcom/test/B;->bar()");

        A.foo(); // calls B.bar()

        Instrumentation.callOriginalMethod(A_foo); // calls A.foo()
    }
}
```

Fig. 2. Example illustrating the functionality of the instrumentation library

to ensure instrumentation of security-sensitive methods before their execution, we create an application class that becomes the superclass of the existing application class[1]. Our new class contains a static initializer, which is the very first code executed upon application startup. The initializer loads our native C-library using `System.loadLibrary()`.

Invocations of security-critical methods do *not* need to be rewritten statically. Instead, we use Java Native Interface (JNI) calls at runtime to replace the references to each of the functions to be monitored. More precisely, we call the JNI method `GetMethodID()` which takes a method's signature, and returns a pointer to the internal data structure describing that method. This data structure contains a reference to the bytecode instructions associated with the method, as well as metadata such as the method's argument types or the number of registers. In order to redirect the control flow to our guard method, we overwrite the reference to the instructions such that it points to the instructions of the security guard's method instead. Additionally, we adjust the intercepted method's metadata to be compatible with the guard method's code. In particular, we adjust the number of registers to the guard method's number of registers. This approach works for pure Java methods as well as methods with a native implementation.

We illustrate how to replace a method using the functionality provided by our instrumentation library in Figure 2. Calling `Instrumentation.replaceMethod()` replaces the instruction reference of method `foo()` of class `com.test.A` with the reference to the instructions of method `bar()` of class `com.test.B`. It returns the original reference, which we store in a variable `A_foo`. Therefore, subsequently calling `A.foo()` will invoke `B.bar()` instead. The original method can still be invoked by `Instrumentation.callOriginalMethod(A_foo)`. Note that the handle `A_foo` will be a secret of the security policy in practice, therefore the original method can not be invoked directly by the instrumented app.

Our approach relies only on the layout of Dalvik's internal data structure for methods, which has not changed since the initial version of Android. However,

---

[1] In case no application class exists, we register our class as the application class.

Table 1. Runtime comparison with micro-benchmarks for normal function calls and guarded function calls with policies disabled as well as the introduced runtime overhead

Function Call	Original Call	Guarded Call	Overhead
Socket-><init>()	0.0186ms	0.0212 ms	21.4%
ContentResolver->query()	19.5229 ms	19.4987 ms	0.8%
Camera->open()	74.498 ms	79.476 ms	6.4%

our instrumentation system could easily be adapted if the layout were to change in future versions of Android.

We are not aware of any possibility to bypass or disable our instrumentation in Java code, as this code is strongly typed. It can even handle cases like reflection or externally loaded libraries, which have not been instrumented. However, native code could potentially alter the references we modified, but it wouldn't know the original references, as our native code executes first. Native code could also modify the guard's bytecode instructions or data structures, which is out of the scope of our approach.

4   Evaluation

In the following we present the results of our experimental evaluation. We measure the performance overhead of our call diversion approach through several micro-benchmarks (cf. Table 1.) All benchmarks have been executed on a Google Galaxy Nexus smartphone running Android version 4.1.1 (Jelly Bean). The smartphone has a dual-core 1.2 GHz ARM CPU from Texas Instruments (OMAP 4460) and 1GB of RAM. Our techniques require *no* custom firmware, which allows widespread deployment. We envision a instrumentation process similar to our previous work [1], where a third-party app can be rewritten directly on the phone. The rewriting process adds code to load the policy classes and executes native code that modifies the references to methods that need to be monitored.

For the evaluation of the runtime overhead we chose to conduct time measurements on three method calls with different runtime complexity, namely Socket-><init>(), ContentResolver->query(), and Camera->open(). We measured time using the System->nanoTime() function. One measurement cycle consists of x iterations over the particular function call inside a loop, where $x = 25$ for Camera->open(), $x = 500$ for ContentResolver->query(), and $x = 10000$ for Socket-><init>(). We executed each cycle 10 times per benchmark. Table 1 reports the median runtime for the original function, for the rewritten function with disabled policies (i.e., we directly call the original function), and, finally, the runtime overhead in percent. We do not report the overhead with enabled policies as this would result in negative overhead as the original methods would not be executed.

During the evaluation we found that in a few cases the monitored calls were faster than the original calls, even though we explicitly invoke garbage collection before each cycle to minimize its distortion. These cases are clearly outliers, possibly due to the operating system's scheduling strategies for other apps running on the same phone. The reported median overhead abstracts from such effects and is always positive. While the relative overhead may seem high, the absolute value is almost negligible and does not adversely affect the application's performance, in particular as any realistic program only invokes a limited number of security-sensitive methods. The micro-benchmarks give a worst-case approximation of the overhead incurred by a program that would only invoke protected functionality.

5 Related Work

In order to overcome the limitations regarding Android's security system, researchers have proposed several approaches, most of which require modifications to the Android platform. Nauman et al. [11] present a modification of the Android software stack called *Apex* that enables dynamic permission revocations. Conti et al. [2] go one step further with *CRePE* that integrates a context-related policy enforcement mechanism into the Android software stack. Fragkaki et al [6] recently presented an external reference monitor approach to enforce coarse grained secrecy and integrity policies called *SORBET*. In contrast to all these approaches, our intention was to be able to deploy the system to unmodified stock Android phones. The major drawback of modifying the firmware and platform code is that it requires rooting the device, which may void the user's warranty. Besides, there is no general Android system but a plethora of vendor-specific variants that would need to be supported and maintained across OS updates. Moreover, a bug in the implementation of the new security features may be exploited and thus void all the benefits of the approach. Finally, most users typically lack the expertise to conduct firmware modifications and therefore abstain from installing modified Android versions on their phone.

Recently, several researchers proposed security frameworks for third party apps that do not rely on modifying the stock Android firmware. A concurrent approach called Aurasium [13] rewrites low-level function pointers of the *libc* library in order to intercept interactions between the application and the OS. Most of the functionality that is protected by Android's permission system depends on such system calls and, thus, can be prevented at this level. Their approach can even detect an application that tries to perform security-relevant operations directly from native code, however, only as long as it does not re-implement the *libc* functionality. However, the parameters of the original Java requests need to be recovered from the system calls' low-level byte arrays in order to differentiate security-relevant from benign requests, which may be error-prone and break in the next version of Android at Google's discretion. Similarly, mock return values are difficult to inject at this low level. In contrast, we designed our system to intercept high-level Java calls, which allows for more flexible policies. In particular, we are able to inject arbitrary mock return values, e.g., a proxy object that only gives access to certain data if a policy prohibits to execute a method. Furthermore, we are able to intercept security-relevant methods that do not depend on the libc library. While the idea of modifying function pointers for IRM has been exemplified by Aurasium, our technique extends this idea to redirecting control flow in a virtual machine.

Another recent research [9] called Dr. Android and Mr. Hide places the reference monitor into a separate application. This allows to remove all permissions from the monitored application, as all calls to sensitive functionality are done in the monitoring app. It is fail-safe by default as it prevents both reflection and native code from executing such functionality. This approach, however, has some major drawbacks: If a security policy depends on state of the monitored application, it incurs high complexity as all relevant data must be marshaled to the monitor. Besides, the monitor may not yet be initialized when the app attempts to perform security-relevant operations. The marshaling of arguments and return values may also degrade application performance. Finally, a bug in the monitor may again lead to privilege escalation, as the monitor delegates the original call and must therefore have the permissions of all monitored apps.

Davis et al. present with *I-ARM-DROID* [3] another inline reference monitor based approach to enforce security policies. Their approach does not allow the instrumentation of applications on the phone so far, however, it supports to instrument calls to any Java method and covers reflective JAVA calls.

In our previous work [1] we presented AppGuard, a practical approach for enforcing fine-grained and stateful security policies on Android apps without requiring changes to the stock Android firmware. It provides a quick mitigation technique for upcoming security vulnerabilities and allows on-the-phone instrumentation of apps. Besides the instrumentation of any Java method and efficient handling of reflective method calls, AppGuard has the smallest runtime overhead of all mentioned approaches. AppGuard proved its feasibility in a real world setting by enforcing several policies on real apps from Google Play. It has already been downloaded more than 500,000 times and was recently asked to join the invite-only Samsung store.

Our idea builds on the notion of diverting the control flow for binary applications as exemplified by, e.g., the Seccomp sandbox on Linux platforms [7], or Detours [8] on Windows NT.

6   Conclusions

We presented an efficient new approach to inline reference monitoring for Android apps. Our call diversion approach follows the idea of callee-site rewriting and heavily reduces the number of changes that have to be performed during app instrumentation. Furthermore, it reduces the runtime of the inlining process and facilitates on-the-phone instrumentation. Our approach allows for dynamic updates of security policies as only references to bytecode need to be changed. We demonstrated the feasibility of the approach through an experimental evaluation and are now working on merging this approach with our prior work on AppGuard.

Acknowledgement. We thank Sven Obser from Backes SRT for his assistance with the experiments. Further, we would like to thank the anonymous reviewers for their comments. This research was supported by the German Federal Ministry of Education and Research (BMBF) within the Center for IT-Security, Privacy and Accountability (CISPA) at Saarland University.

References

1. Backes, M., Gerling, S., Hammer, C., Maffei, M., von Styp-Rekowsky, P.: App-guard - real-time policy enforcement for third-party applications. Tech. Rep. A/02/2012, Saarland University, Computer Science (July 2012), http://www.infsec.cs.uni-saarland.de/projects/appguard/android_irm.pdf
2. Conti, M., Nguyen, V.T.N., Crispo, B.: CRePE: Context-Related Policy Enforce-ment for Android. In: Burmester, M., Tsudik, G., Magliveras, S., Ilić, I. (eds.) ISC 2010. LNCS, vol. 6531, pp. 331–345. Springer, Heidelberg (2011)
3. Davis, B., Sanders, B., Khodaverdian, A., Chen, H.: I-ARM-Droid: A rewriting framework for in-app reference monitors for android applications. In: Mobile Secu-rity Technologies 2012, MoST 2012 (2012)
4. Desmet, L., Joosen, W., Massacci, F., Naliuka, K., Philippaerts, P., Piessens, F., Vanoverberghe, D.: The s3ms.net run time monitor. Electron. Notes Theor. Com-put. Sci. 253(5), 153–159 (2009)
5. Erlingsson, Ú., Schneider, F.B.: Irm enforcement of java stack inspection. In: Proc. 2002 IEEE Symposium on Security and Privacy (Oakland 2002), pp. 246–255 (2000)
6. Fragkaki, E., Bauer, L., Jia, L., Swasey, D.: Modeling and Enhancing Android's Permission System. In: Foresti, S., Yung, M., Martinelli, F. (eds.) ESORICS 2012. LNCS, vol. 7459, pp. 1–18. Springer, Heidelberg (2012)
7. Google Seccomp sandbox for Linux, http://code.google.com/p/seccompsandbox/
8. Hunt, G., Brubacher, D.: Detours: binary interception of Win32 functions. In: Pro-ceedings of the 3rd Conference on USENIX Windows NT Symposium, WINSYM 1999. USENIX Association, Berkeley (1999)
9. Jeon, J., Micinski, K.K., Vaughan, J., Fogel, A., Reddy, N., Foster, J., Millstein, T.: Dr. Android and Mr. Hide: Fine-grained permissions in android applications. In: 2012 ACM CCS Workshop on Security and Privacy in Smartphones and Mobile Devices, SPSM (2012)
10. Ligatti, J., Bauer, L., Walker, D.: Edit automata: Enforcement mechanisms for run-time security policies. International Journal of Information Security 4(1-2), 2–16 (2005)
11. Nauman, M., Khan, S., Zhang, X.: Apex: Extending android permission model and enforcement with user-defined runtime constraints. In: Proc. 5th ACM Sym-posium on Information, Computer and Communication Security, ASIACCS 2010, pp. 328–332 (2010)
12. Schneider, F.B.: Enforceable security policies. ACM Transactions on Information and System Security 3(1), 30–50 (2000)
13. Xu, R., Saïdi, H., Anderson, R.: Aurasium – practical policy enforcement for an-droid applications. In: Proc. 21st USENIX Security Symposium (2012)

Towards Unified Authorization for Android

Michael J. May[1],[*] and Karthikeyan Bhargavan[2]

[1] Kinneret College on the Sea of Galilee, DN Emek Hayarden 15132, Israel
[2] INRIA

Abstract. Android applications that manage sensitive data such as email and files downloaded from cloud storage services need to protect their data from malware installed on the phone. While prior security analyses have focused on protecting system data such as GPS locations from malware, not much attention has been given to the protection of application data. We show that many popular commercial applications incorrectly use Android authorization mechanisms leading to attacks that steal sensitive data. We argue that formal verification of application behaviors can reveal such errors and we present a formal model in ProVerif that accounts for a variety of Android authorization mechanisms and system services. We write models for four popular applications and analyze them with ProVerif to point out attacks. As a countermeasure, we propose Authzoid, a sample standalone application that lets applications define authorization policies and enforces them on their behalf.

1 Introduction

The Android operating system seeks to foster a rich ecosystem of third-party applications. Users may download apps from reputable stores managed by Google and Amazon or directly from app developers.[1] This leaves users vulnerable to malware masquerading as genuine apps. Consequently, Android provides strong runtime isolation, running each application process in a separate Dalvik virtual machine, and giving each a private storage area. Isolated applications may still share files and data, for example using external storage or using an inter-app messaging mechanism called *intents*. While some apps freely share and collaborate with others, those holding sensitive data are tempered by the need for security and integrity. Android therefore provides authorization mechanisms which let an app control which other apps, if any, can read or write its data.

System Permissions. Android protects its system resources through *permissions* which are granted by the user at installation time and accompany the app throughout its lifetime. The Android SDK defines about 130 built-in permissions of which some forty are *signature/system* permissions and are reserved for the operating system or apps installed by the device manufacturer [28]. The rest can be requested by an app in its *application manifest*, an XML file which

---

[*] Work performed while visiting INRIA.
[1] It is estimated that the average Android phone in 2012 has 32 apps installed [24].

J. Jürjens, B. Livshits, and R. Scandariato (Eds.): ESSoS 2013, LNCS 7781, pp. 42–57, 2013.

is prepared by the developer and stored in its application package (APK) file. When an app attempts to access a system API function at run time, Android first checks if the requestor has the required permission. If it doesn't, a security exception is thrown.

Some examples of regular permissions are: READ_EXTERNAL_STORAGE and WRITE_EXTERNAL_STORAGE permissions to read or write to the phone's shared disk (referred to informally as the *SD card* since its default mount point is /mnt/sdcard/ [21]), INTERNET to open network sockets, and READ_LOGS[2] to access the system log. Some examples of system permissions are: INSTALL_PACKAGES to install new applications, BRICK to disable the phone completely, and DELETE_CACHE_FILES to clear the cache directories of other apps.

Application-level Authorization. In addition to install-time permissions, Android provides a variety of other authorization mechanisms. Activities can filter which intents will be directed to them based on the intent's content or requested action. Content providers and services can be made available only to applications which have certain permissions. Authorization may seem seamless to the user, but due to the variety of tools available and the details of the OS, it can be technically messy and sometimes can even be bypassed.

Consider, for example, a user who installs Dropbox (www.dropbox.com), uses it to download a PDF file from the cloud, and opens it with Adobe Reader (adobe.com/products/reader.html).[3] The user would assume that during the transaction Adobe Reader got temporary read access to the file and nothing more. As we discuss later, that is not the case at all: Adobe Reader and (up until API level 16) *all* of the applications on the phone can read the file indefinitely afterwards. Many can modify it too.

Our Contribution. There are many ways in which applications get authorization wrong or fail to enforce authorization properly. They fail primarily because they don't define the policy they are trying to enforce and (likely) didn't use a full model of the environment during testing. Proper modeling of the environment and the application's behavior would reveal attacks on the authorization mechanism.

Our contribution in this work is threefold. First, we present a unified picture of the Android authorization tools, something not previously presented in a single work. Second, we show how many popular sharing applications on the market fail to get authorization right and publish a formal model of the Android authorization tools and environment which allows us to reveal attacks. We publish the model so that others can use and extend it to test their apps. Third, we present Authzoid, a sample authorization app which properly implements the authorization tools that the apps got wrong. The code can act as a source code module to be included as is or as a starting point for developers who want to get authorization right. Both code and model are found at: http://prosecco.gforge.inria.fr/Essos/pv/.

---

[2] Changed to signature/system/development permission in API level 16.
[3] The most popular PDF reader on Google Play as of Oct 2012.

The rest of this paper is organized as follows. Section 2 explains the Android authorization tools. Section 3 discusses our attacker model, gives technical descriptions of the sharing applications surveyed, and explains the attacks against them. Section 4 contains our formal model for Android authorization tools, environment, and the applications studied. Section 5 discusses the Authzoid app and its major features. Section 6 contains related work and Section 7 concludes.

2   Authorization for Android Applications

Android applications are composed of four kinds of run time entities:

Activities correspond to windows and allow for user interaction via a GUI.
Content Providers provide SQL-like interfaces to queryable data.
Broadcast Receivers listen for broadcast messages from other application
  components, the operating system, or other applications.
Services run in the background and provide long term functionality without
  providing a user interface.

Applications exchange messages via intents which contain a URI data field and strings, URIs, or key-value pairs in *extra* fields. They are routed by an Android component called *Binder* between the run time entities. During routing, the *Activity Manager* writes the action, sender, recipient, and data field (but not the extra fields) to the log.

Each runtime entity may use a variety of authorization mechanisms to control access to its data and functionality. In the rest of this section we review the five authorization mechanisms available and explain their usage, strengths, and weaknesses. For each mechanism, we give an example of how it is used in a popular application currently available from the Android Market. Some apps use a combination of the mechanisms below to enable a variety of user policies.

Android Permissions. Android's SDK includes about 130 permissions, but an app may extend them with its own permissions by adding them to its manifest file. Apps can use permissions to enforce authorization in one of two ways.

First, content providers, activities, services, and broadcast receivers can specify that only applications with a certain permission may access them. For activities and services, this prevents applications without the permission from invoking or binding to them. For content providers, separate read and write permissions may be given. For broadcast receivers, it prevents the delivery of broadcasts from apps without the permission. The filtering is done automatically by Binder based on the app's manifest file (`android:permission` for activities, services, and broadcast receivers and `android:readPermission` and `android:writePermission` for content providers) or as defined programmatically if they are configured in code.

Example: K-9 Email (`code.google.com/p/k9mail/`) uses the custom permissions `com.fsck.k9.permission.READ_ATTACHMENT` to protect its attachments content provider. Its email messages content provider protects read access with `READ_MESSAGES` and write access with `DELETE_MESSAGES`. GMail (`gmail.com`) and Yahoo! Mail (`mail.yahoo.com`) (discussed below) use a similar strategy.

Second, apps can use the system API to discover whether it, the app which called it, or another app has a particular permission (using `checkPermission()` or `checkCallingPermission()`), regardless of its type. They can then make programmatic decisions based on the results.

Example: Some plug-in libraries (*ex.* ACRA (`code.google.com/p/acra/`)) programmatically investigate which permissions are available to their host applications before attempting actions which require particular permissions (*ex.* reading the log and sending internet data).

URI Permissions The content provider read and write manifest permissions give blanket read and write access. Alternatively, a content provider can give specific read or write query access to a single `content` URI under its authority. The URI permission can be granted programmatically using `grantUriPermission()` or by sending an intent to the recipient with the `FLAG_GRANT_READ_URI_PERMISSION` or `FLAG_GRANT_WRITE_URI_PERMISSION` flags set. URI permissions can be delegated by recipients. Intent-granted URI permissions are valid until the recipient app closes or is killed. Programmatically-granted ones are valid until revoked using `revokeUriPermission()`.

Binder enforces URI permissions by tracking the grants, revokes, and intents sent, so the URI does not need to be secret. Also, since intents can be routed by capability, the sender may not know which app received the permission.

Depending on whether the `content` URI refers to a database row, a file, or both, the recipient can use a *content resolver* request to read or write the corresponding rows or file. If the URI is opened as a file using `openFile()`, the content provider returns an open file descriptor for it and Binder assigns ownership of it to the recipient.

Example: Users can open an attachment from K-9 Email with an external viewer. When this happens, K-9 Email sends an intent to the viewer with a `content` URI for the attachment and the URI read permission flag set. The recipient can then use a content resolver to resolve the URI to an open file descriptor. GMail and Yahoo! Mail employ a similar strategy.

The use of open file descriptors leads to some technical inconveniences. First, since a file descriptor is a hard link to the file and is owned by the recipient, the sender can't close the file descriptor or delete the file until the recipient closes it. Second, if two application hold open file descriptors for the same file (*i.e.* they both requested the same URI), they cause read/read and read/write conflicts and race conditions. Third, only a few classes in the Java file API support file descriptors, making it impossible to perform random access reads or writes to the file and making rewinding difficult. Because of these issues, some applications immediately make local copies of files passed to them by URI (*ex.* Adobe Reader) or don't enable updates to such files (*ex.* Jota Text Editor (`sites.google.com/site/aquamarinepandora/home/jota-text-editor`)).

Private Storage. Every Android application is given its own user name, group name, and home directory. The home directories are protected by Linux file and directory permissions and by default no app can read or write the home directory of another. Apps can override the default settings to make files or directories world readable, writable, or executable using `setReadable()`, `setWritable()`, and `setExecutable()`. Then any other app can read, write, or execute the files or directories. If an app makes a file world readable in order to share it, it may include a long random string in the path to make it hard for unintended apps to guess the path name. This technique turns the path name into a secret, so the app must ensure that only the intended recipient gets the path name.

Example: Unlike most apps which keep all files in private storage private, Google Drive (`drive.google.com`) (discussed below) selectively sets path read and execute permissions to enable others to read files in its private storage.

External Storage (SD Card). Most Android devices have a shared storage space for files or data (the SD card). Read and write access to the SD card require the permissions mentioned above. Many applications (*ex.* the camera, the default browser's downloads folder) use the SD card for storing files that are either too large to keep in private storage or that are meant to be available for other apps to use. Aside from the read and write permissions, Android does not enforce access control on the SD card, so any application can read, write, or delete any file on it. Authorization can be enforced on the SD card using encryption or message authentication codes (MAC).

Example: The password storage app 1Password (`agilebits.com/onepassword`) stores its encrypted password database on external storage. It doesn't share passwords directly with other apps, instead using the clipboard to copy and paste passwords. Encryption protects the contents of the password database and MACs protect its integrity.

Web Sharing. Some apps place data to be shared on a public web site and send the URL for the data to another app via an intent. Often the URL contains a long random string to make it difficult to guess, turning the URL into a secret. Another option is to protect the URL using web-based application or user authentication such as OAuth [14].

Example: MyTracks (`www.google.com/mobile/mytracks/`) uses GPS information to track where the device has gone, including distance traveled, speed, and elevation change. When sharing a "track" from MyTracks with another app, it first uploads a custom map to Google Maps and then sends a web URL with a long random part to the recipient via an intent.

Summary. The authorization tools listed show the variety of mechanisms available. It's not clear if any or all of the tools are sufficient to achieve a satisfactory
authorization policy. The applications that we study in the next section use different combinations of the tools to enforce authorization, but each suffer from
attacks and weaknesses that demonstrate that using them correctly is not simple. In some, the tools are used incorrectly; in others, features of the Android
environment defeat the app's authorization policy.

3   Applications and Attacks

To investigate how popular applications use the authorization tools of Section
2 to enforce their security goals, we study four apps: two Email clients and two
Cloud File Storage applications. We explain each application's authorization
mechanism and explain how an attacker may defeat it.

3.1   Authorization Goals and Attacker Model

Since the apps we examine don't specify authorization policies in their documentation, we define a minimal one for the purposes of our study. Our minimal
policy contains just one confidentiality rule and one integrity rule:

Confidentiality. An app may read a file only if it owns it or if the owner and
   the user have authorized the reading.
Integrity. An app may modify a file only if it owns it or if the owner and the
   user have authorized the modification.

The policy can be enforced by many authorization mechanisms, including those
listed in Section 2.

Regarding the attacker model, we first assume that the Android protection mechanisms are enforced according to their specification (*i.e.* the phone
isn't "rooted", giving arbitrary power to an app). Next, we make the same
assumption that Android does regarding app isolation: that apps are mutually suspicious. The attacker is assumed to be (1) installed on the phone, (2)
capable of performing polynomial time programmatic tasks, and (3) in pos-
session of a set of authorization-related permissions that seem may seem in-
nocuous: READ_EXTERNAL_STORAGE, WRITE_EXTERNAL_STORAGE (51% of popular
applications request it), READ_LOGS (6% of popular applications request it[4]),
and INTERNET (77% of popular applications request it) [10]. Any series of actions which such an attacker can take to contravene the authorization policy
defined above is an attack.

---

[4] As of Oct 2011, READ_LOGS was the ninth most popular dangerous Android permission requested. In API level 16, it was converted to a system/signature/development
permission, so access to it on the most recent devices is significantly reduced.

3.2   Study of Sharing Applications

We now consider four popular Android applications which enforce authorization using the mechanisms defined in Section 2. The applications are chosen because they illustrate the use of a variety of mechanisms and are representative of classes of apps.

GMail downloads attachments to its private storage area and manages them via a content provider which is protected by custom permissions `READ_GMAIL` and `WRITE_GMAIL`. The permissions are signature level permissions, so only Google applications can request them [7]. GMail allows the user to open an attachment using an outside document viewer by sending an intent containing a URI read permission. Applications which behave similar to GMail include the built in Android Email application and K-9 Mail.

Attacks: The use of a protected content provider ensures that only applications sent the URI permission can read the file. However, some recipient viewers immediately make a copy of any file sent to them by URI. For example, Adobe Reader copies any file it shows to the SD card in the downloads directory, making it readable by an attacker (confidentiality).

Yahoo! Mail has a content provider which is protected by a custom signature permission (`com.yahoo.mobile.client.android.permissions.YAHOO_INTER_APP`). Yahoo! lets the user open attachments using URI permissions, just like GMail. However, downloaded attachments are stored on the SD card, so they are readable by any application with `READ_EXTERNAL_STORAGE`. The MailDroid (`groups.google.com/group/maildroid`) application behaves similarly.

Attacks: Since Yahoo! stores all downloaded attachments on the SD card, an attacker can read them (confidentiality). The application does not check for downloaded file integrity, so once on the SD card they may be modified by an attacker as well (integrity).

Google Drive offers two mechanisms for sharing files on the phone.

First, the on-phone app lists the files and directories on the cloud and downloads one when the user requests to view or share it. Files can't be updated on the phone. A downloaded file is placed in a new, randomly named directory in a document cache directory located in Google Drive's private storage (`/data/data/com.google.android.apps.docs/cache/`). The new directory contains just one file and is world readable and executable. The file is made world readable and all the directories above the randomly named directory are world executable, letting any app open the file, but not list the directory names under `cache`. The file path is sent to the target as the `data` field of an intent.

Second, apps can access Google Drive via a web service interface which is protected by SSL and OAuth 2.0. An app receives an API identifier which it can use to obtain file read and write tokens. An app can download files via their names or identifiers and send updates back over the web.

Attacks: With respect to the on-phone app, the directory path is a secret since any application which knows it can open the file. Activity Manager, however, prints the `data` fields of all intents to the log, so the path is printed as well. An attacker which has `READ_LOGS` permission can discover the path and read the file (confidentiality).

Dropbox lists the files and directories in the cloud and downloads them on demand. The files are stored on the SD card in a directory called `scratch`. When sharing a file, it is first stored in the `scratch` directory and then the path and filename are sent via an intent.

Downloaded files can be opened for reading and editing, but are not checked for integrity after downloading. When opening a file, the user chooses which file viewer to use; if the viewer saves a new version, it is uploaded to the cloud. Saves are monitored until the authorized viewer closes or loses focus. Dropbox ignores saves by other applications, even when an authorized viewer is working.

When sharing a file for attachment to an email, the file is uploaded (if necessary) and a web URL is provided in the intent as an extra. The URL provides read only access to the file and includes a random string to make guessing harder.

Attacks: Since downloaded files are stored on the SD card, they are readable by an attacker (confidentiality). Unauthorized saves are not automatically uploaded to the cloud, but since there is no integrity check, an attacker can tamper with a file and subsequent views of the file on the phone will show the tampered version. If a viewer unknowingly saves a tampered version, the attacker's modifications will reach the cloud (integrity).

3.3 Discussion

Our study of four popular and well-regarded applications illustrates the difficulty in getting even a simple authorization policy right. Many applications place sensitive data on public external storage. Some use unguessable directory names in private storage, but these names may leak into the shared system log. Still others may themselves correctly implement access control, but may be let down by the applications with which they share files.

Simple technical tricks aren't sufficient against a dedicated adversary. Wuala (`wuala.com`) tries to place shared files on the SD card for only a few seconds. However, due to Android's application life cycle, a malicious app can monitor the SD card and breach confidentiality during that gap. Boxcryptor (`boxcryptor.com`) encrypts files on the SD card and decrypts them just in time for the recipient. Such uses of encryption are limited by key management, that is, how to securely transfer a secret key to the recipient. Google Drive's example shows that transferring secret keys by intent is not always secure. Keys derived from passphrases are hard to keep secret, as shown by Belenko and Sklyarov [3]. Even if encryption keys are shared securely, file storage applications often misuse encryption and integrity algorithms and expose their plaintext to attackers [4].

We advocate a unified comprehensive approach to the implementation of application-level authorization. Rather than suggest point-fixes to prevent specific attacks, we show how to write formal models that precisely capture authorization policies and relevant parts of the execution environment. By automatically analyzing such models, we can both find attacks and gain confidence in the mechanisms used to enforce the policy.

4   Formal Model

As shown above, implementing even a minimal authorization policy requires an analysis of the authorization tools as well as the environment. Modeling can help such an analysis by including relevant parts of the Android authorization tools and the operating system. Developers can then create a model of their application, run it inside the Android model, and use automated tools to discover attacks. In this section we describe the building of such a model using ProVerif [5]. We show illustrative snippets of its parts: (a) the authorization policy, (b) the Android authorization tools, (c) parts of the Android OS, and (d) the sharing application. We then use the model to discover attacks in the models of the applications surveyed. ProVerif is well suited for our needs since (1) it enables the definition of authorization policies using Horn clauses and communication using the applied pi calculus; (2) it can model enforcement mechanisms that use secret and fresh file or path names and cryptography; and (3) it lets us analyze the models against an unbounded adversary.

Policy Language. The snippet below implements the minimal authorization policy from Section 3.1. It allows an app to read or write files only if it is the file's owner or if it receives authorization from the owner and the user. Lines 1–2 are a horn clause saying that if an application (a1) and a user (u) own a resource (r), then a1 is authorized to read r. Lines 3–4 enable another application (a2) to receive read authorization from the owners (a1 and u). The parallel write rules are omitted due to space considerations.

```
1  clauses forall u:Principal, a1:appid, a2:appid, r:resource;
2      owners(r, u, a1) -> readAuthorized(a1, r).
3  clauses forall u:Principal, a1:appid, a2:appid, r:resource;
4      owners(r, u, a1) && userAuthorizedRead(u, a2, r) -> readAuthorized(a2, r).
```

Android Authorization Tools. We implement the following Android authorization tools:

Permissions are included via a `androidPerm` type which is populated with permissions that can be granted by the user during installation.
URI Permissions are included via a `uri` type which refers to a file resource. Resolution is modeled using a lookup table of `uri` and `resource` pairs.

Private Storage is modeled by using a `path` type which refers to a location only accessible by the owner. If the path is declared world readable, writable, or executable, others can access it too. Fresh path names may be world readable, but only can be accessed if the requestor knows the path's name.

SD Card is modeled as a file system process which enables the storage or retrieval of objects based on `path` and `filename` objects stored in a lookup table.

Web Sharing is parallel to private storage, but without the need for setting path permissions.

The following snippet shows how file and log read permissions are handled (parallel file write clauses are elided). Lines 5–7 define the Android permission type, the permission to read the SD card (externalRead), and the permission to read the log (logRead). Lines 8–9 allow an application a to read a file with name f, path p, and any file and path permissions fp and pp if (1) a has the external read permission and (2) the file is on the SD card. Lines 10–11 allow an application a to read all files in its own private space (private(a)). Lines 12–14 allow an application a to read a file in another application o's private space if its path permissions (pp) are set to world executable (isWorldExecutable) and its file permissions (fp) are set to world readable (isWorldReadable). Line 15 allows an application a to read the log if it has logRead.

```
5    type androidPerm.
6    fun externalRead() : androidPerm.
7    fun logRead() : androidPerm.
8    clauses forall a:appid, l:location, p:path, f:filename, pp:filePerms, fp:filePerms;
9      hasPermission(a, externalRead()) -> canReadFile(a, sdcard(), p, pp, f, fp).
10   clauses forall a:appid, l:location, p:path, f:filename, pp:filePerms, fp:filePerms;
11     canReadFile(a, private(a), p, pp, f, fp).
12   clauses forall o:appid, a:appid, p:path, f:filename, pp:filePerms, fp:filePerms;
13     isWorldExecutable(pp) && isWorldReadable(fp) ->
14     canReadFile(a, private(o), p, pp, f, fp).
15   clauses forall a:appid; hasPermission(a, logRead()) -> canReadLog(a).
```

Android OS Elements. We include processes for the following authorization related Android processes:

File System process which enables applications to read, write, and list files on the SD card based on path and file name. The file system allows access to files in private storage by the owner and by others which know the path name if the permissions are set correctly.

Content Provider process which enables applications to resolve URIs and thereby read or write files which they refer to.

Binder process which handles the granting of URI permissions, both from the owner and via delegation. Binder writes entries in the log.

Log process which gives permission-based read and write access to the log.

Permission Granting process which enables the user to grant permissions to
processes.

The following snippet shows three parts of file system's code in the model. Lines
16–17 listen for file read requests (readFile) and check the application is registered. Lines 18–19 retrieve the file based on its location (l), path (p), and file
name (f), check if a is able to read it, and return it to a if it is able. Lines 20–22
listen for requests to list the files in a directory path. Line 23 allows it if the
path is world readable. Lines 24–27 allow an application to list all files on the
SD card if the requestor has externalRead permission.

```
16  let FileSystem() = (!in (filesystem,readFile(a, l, p, f));
17     get apps(=a) in
18     get files(=l, =p, pp, =f, fp, r) in
19     if canReadFile(a, l, p, pp, f, fp) then out(return(a), r))
20  | (!in (filesystem,listFile(a, l, p));
21     get apps(=a) in
22     get files(=l, =p, pp, f, fp, r) in
23     if isWorldReadable(pp) then out (return(a), f))
24  | (!in (filesystem,listSDCard(a));
25     get apps(=a) in
26     get files(=sdcard(), p, pp, f, fp, r) in
27     if hasPermission(a, externalRead()) then out (return(a), (p, f))).
```

Testing Application. We implement a single process for each sharing application.
The process registers the application, specifies how files are added, and specifies
how files are shared with other applications ("open with").

The following snippet shows the Dropbox application. Line 28 defines the
Dropbox application id as private (not known to the attacker initially). Line
29 is the header for the process. Lines 30–31 register Dropbox in the applications table (apps, definition elided) and publish the name of its private storage
(private(dropbox)) and its web storage (web(dropbox)) by sending their values
on the free channel pub (definitions of private, web, and pub are elided). This
simulates an attacker knowing the application's root directory and web domain,
but not knowing the paths below them where files are found. Lines 32–34 define
a new file's contents (r), its file name f, and its path p; assigns ownership of the
file to Dropbox (line 33, using the assume function which is elided); and inserts
it in the files table (definition elided) on the web (web(dropbox)) where it stays
until downloaded. The noPerms() terms are used to model file and path read,
write, and execute permissions. Lines 35–40 model a user opening a file. Lines
35–36 receive an openWith command and download a file from Dropbox's web
space (path p, path permissions pp, file name f, file permissions fp, file contents
r). Line 37–38 select an application a and authorize it to read. Line 39 stores
the file on the SD card in the files table with no explicit file or path permissions
(noPerms). Line 40 returns the path and file name to the requestor by an explicit
intent (explicitintent(a)).

```
28   free dropbox : appid [private].
29   let Dropbox(u:Principal) =
30     (insert apps(dropbox);
31     out (pub, (private(dropbox), web(dropbox))))
32   | (!new r:resource; new f:filename; new p:path;
33     if assume(owners(r, u, dropbox)) then
34     insert files(web(dropbox), p, noPerms(), f, noPerms(), r))
35   | (!in (openWith, ());
36     get files(=web(dropbox), p, pp, f, fp, r) in
37     get apps(a) in
38     if assume(userAuthorizedRead(user, a, r)) then
39     insert files(sdcard(), p, noPerms(), f, noPerms(), r); (*readable by attacker*)
40     out (explicitintent(a), (p, f))).
```

Discovery of Attacks. By combining the testing application's model with the environment and authorization code, we can check two types of queries in ProVerif:

1. Checks that proper authorization is reachable. ProVerif should show traces by which a user can properly authorize an app to read and write a file.
2. Checks that an attacker can't read or write files without proper authorization.

The first queries check that authorization is possible under the model. We expect to see traces of the sort: "The file is readable by the attacker if the user has sent a read URI permission to the attacker" or "A file in private storage is readable by the attacker if the application makes the file world readable, the path world executable, and sends an intent to the attacker with the path to file." Those represent valid authorization paths in the model.

The second queries ensure that there are no other ways to read or write files aside from the authorization paths defined. If ProVerif finds any such paths, they are attacks. For example, ProVerif points out that line 39 above leaks the file to the attacker since it is allowed to read files on the SD card.

5 Authzoid Implementation

As shown above, the proper use of the authorization tools in Android requires careful design and analysis. In this section we describe our implementation of Authzoid, an app that lets file owners define authorization policies and then enforces them on their behalf. Authzoid uses the authorization tools explained in Section 2 to enable a wide variety of policies, including ones far richer than the minimal policy defined in Section 3.1. Authzoid is useful as a sample implementation of proper authorization, fixing the mistakes of the apps discussed above and can be useful as a starting point for developers who want to get authorization right. Its source code can be found at: http://prosecco.gforge.inria.fr/Essos/pv/.

Authzoid offers three application-facing interfaces: file submission, policy definition, and file retrieval. It manages file versions and authorization checks internally.

Submission Interface. Applications can submit files to Authzoid for storage using an intent with a custom action. The intent can contain a file to share or a content URI to resolve. Files can be submitted as new or as updates to existing files.

If submitted as new, Authzoid retrieves the name of the submitting application via the Android API and stores it in its private storage area. A private database indexes the files by their original file name or URI and submitting application. The new file is assigned a new version number which is returned to the submitter. The submitter may optionally include a permission as a string extra. If included, any application with the given permission may later read or modify the file (see below).

If submitted as an update, the file must be accompanied by the name of the owner, the file's original path and file name, and a version number which indicates the last version of the file the submitter saw. Authzoid first checks if the submitter is authorized to update the file (see below). If not, an authorization failure message is returned. If the update is authorized, but the version number submitted is smaller than the current version number in the database, the update is rejected with an explanatory message. Otherwise, the file is copied in to private storage and the database is updated. Authzoid generates a new version number which it stores in the database and sends it back to the submitter.

Policy Definition. By default, only the application which submitted a file (its owner) can read or write it. Authzoid enables owners to share the file via URI permissions, by adding another app to the file's read and write access control list, or by retrieving a read-only randomized path name (similar to Google Drive). Groups of apps can be added by setting a permission on the file; then any application with the permission can read or write the file.

Retrieval Interface. An app can request a file from Authzoid by sending an intent with owner's name, the file's original path, and name. For each request, Authzoid queries its access control matrix to see if the requestor is authorized to read the file. If the read is approved, Authzoid checks if a copy of the latest version of the file is already in the cache. If not, it generates a new directory under its private `filecache` directory with a 128-bit random name and puts a copy of the file in it. The file and random directory are set to be world readable and the directory is set to be world executable. Whether new or existing, the full file path of the file are returned to the requestor using an intent with the full path in an extra.

When an application resolves a URI using Authzoid's content provider, the content provider makes a new copy of the file, opens a new file descriptor on it, deletes the file using the Java file API, and then returns the file descriptor. This prevents read/read conflicts on the file. Since the file descriptor acts as a hard link, the Android OS will preserve the contents of the file until the recipient closes the file descriptor or is killed.

Folder listings can be requested by sending the owner's name and the path via an intent. Authzoid checks its access control matrix to see if the requestor is the owner or authorized to list the directory. If authorized, a listing of all files and directories in and under the given directory is sent back via an intent as a string array extra.

Authzoid is the first Android app that provides a unified authorization service enabling file sharing between Android apps. Using ProVerif, we verified that Authzoid is secure against the class of attacks captured in our formal model. This should not be interpreted as a formal theorem however, since our model of both Android is abstract and incomplete, and may hide other attacks. Still, our analysis presents a first step towards formal security analysis for Android applications. Our models are public and may be extended for more sophisticated analysis.

6 Related Work

Research on Android's security infrastructure includes studies on how permissions are enforced [17], used [2], and misused or attacked [10,12,18,22]. Some try to secure Android applications against attackers by performing static or dynamic analysis of apps (ex. [16,8,20]). Xu, et al. [29] developed Aurasium, a tool that uses static analysis and code injection to detect or prevent privilege escalation attacks. Like Authzoid, Aurasium does not require modifications to the operating system. Conti, et al. [11] developed CRePE, a system capable of enforcing rule based context aware security policies. Naumann, et al. [23] extended Android permission with custom user defined constraints. None of the above work includes formal analysis or verification.

Research on formalization of the Android stack and API includes Chaudhuri [9] who gave a formal model of a subset of the Android communication system; Enck, et al. [15] who developed TaintDroid to track the flow of sensitive information between Android apps (extended by Shreckling, et al. [25] with more complicated, dynamic run time policies); and Armando, et al. [1] who presented a more complete model of the Android middleware using types.

With respect to formalizations for secure sharing of resources, Blanchet and Chaudhuri [6] developed a formally verified protocol for secure file sharing on untrusted storage (a tool which could be used to secure Android's SD card) and Fragkaki, et al. [19] gave formal typing rules to explain Android's security model. Similar to our work, Fragkaki et al. described Sorbet, a modification to Android which enforces secrecy and integrity properties written by app developers. In contrast, Authzoid is developed to enable the easy specification of authorization policies and relies upon existing Android mechanisms without requiring changes to the operating system.

Since many Android apps are distributed free and make money from in-app ads, work has been done to determine how ad libraries operate and whether they pose privacy or authorization risks. Dietz, et al. [13] developed Quire which enabled advertisers to prevent app based ad fraud. Stevens, et al. [27] investigated the behavior of thirteen ad libraries and showed how their requirements cause app developers to request more permissions than necessary (*permission bloat*). As a remedy to permission bloat, Shekhar, *et al.* [26] implemented AdSplit, a mechanism to separate ad libraries from individual apps.

7   Conclusion

Many Android apps attempt to enforce authorization policies for sharing resources, but fail due to misuse of the Android authorization tools or due to actions by external entities. We can discover authorization attacks by using ProVerif to model a relevant subset of the Android authorization tools and environment and use it to examine the behavior of sharing applications. We also describe Authzoid, an application which lets app developers specify authorization policies for sharing and enforces them using built-in Android tools. Future extensions to Authzoid include work on making an encrypted cache on the SD card and enabling it to proxy OAuth based web sharing.

References

1. Armando, A., Costa, G., Merlo, A.: Formal modeling and reasoning about the android security framework. In: 7th Intl Sym. on Trustworthy Global Computing (2012)
2. Barrera, D., Kayacik, H.G., van Oorschot, P.C., Somayaji, A.: A methodology for empirical analysis of permission-based security models and its application to android. In: 17th ACM Conf. on Computer and Comm. Security, CCS 2010 (2010)
3. Belenko, A., Sklyarov, D.: "Secure Password Managers" and "Military-Grade Encryption" on Smartphones: Oh, Really? Technical report, Elcomsoft Ltd. (2012)
4. Bhargavan, K., Delignat-Lavaud, A.: Web-based attacks on host-proof encrypted storage. In: 6th USENIX Workshop on Offensive Technologies, WOOT 2012 (2012)
5. Blanchet, B.: An efficient cryptographic protocol verifier based on Prolog rules. In: Computer Security Foundations Workshop, CSFW 2001 (2001)
6. Blanchet, B., Chaudhuri, A.: Automated formal analysis of a protocol for secure file sharing on untrusted storage. In: IEEE Sym. on Security and Privacy, SP 2008 (2008)
7. Bray, T.: Recent Android app update prevents third-party apps from using com.google.android.gm.permission.READ_GMAIL. Why? (July 29, 2011), productforums.google.com/d/msg/gmail/XDOC4sw9K7U/8KwuZ10R168J
8. Chan, P.P.F., Hui, L.C.K., Yiu, S.M.: Droidchecker: analyzing android applications for capability leak. In: ACM Conf. on Security and Privacy in Wireless and Mobile Networks, WISEC 2012 (2012)
9. Chaudhuri, A.: Language-based security on android. In: ACM SIGPLAN Fourth Workshop on Programming Languages and Analysis for Security, PLAS 2009 (2009)
10. Chia, P.H., Yamamoto, Y., Asokan, N.: Is this app safe? A large scale study on application permissions and risk signals. In: WWW 2012 (2012)
11. Conti, M., Nguyen, V.T.N., Crispo, B.: CRePE: Context-Related Policy Enforcement for Android. In: Burmester, M., Tsudik, G., Magliveras, S., Ilić, I. (eds.) ISC 2010. LNCS, vol. 6531, pp. 331–345. Springer, Heidelberg (2011)

12. Davi, L., Dmitrienko, A., Sadeghi, A.-R., Winandy, M.: Privilege Escalation Attacks on Android. In: Burmester, M., Tsudik, G., Magliveras, S., Ilić, I. (eds.) ISC 2010. LNCS, vol. 6531, pp. 346–360. Springer, Heidelberg (2011)
13. Dietz, M., Shekhar, S., Pisetsky, Y., Shu, A., Wallach, D.: Quire: Lightweight provenance for smart phone operating systems. In: 20th USENIX Conf. on Security (2011)
14. Hammer-Levy, E. (ed.): The OAuth 2.0 Authorization Protocol. IETF (September 22,.2011), draft-ietf-oauth-v2-22. Work in Progress (Expires March 25, 2012)
15. Enck, W., Gilbert, P., Chun, B., Cox, L., Jung, J., McDaniel, P., Sheth, A.: Taintdroid: an information-flow tracking system for realtime privacy monitoring on smartphones. In: 9th USENIX Conf. on Operating Systems Design and Implementation, OSDI 2010 (2010)
16. Enck, W., Ongtang, M., McDaniel, P.: On lightweight mobile phone application certification. In: 16th ACM Conf. on Computer and Comm. Security, CCS 2009 (2009)
17. Felt, A., Chin, E., Hanna, S., Song, D., Wagner, D.: Android permissions demystified. In: 18th ACM Conf. on Computer and Comm. Security, CCS 2011 (2011)
18. Felt, A., Wang, H., Moshchuk, A., Hanna, S., Chin, E.: Permission re-delegation: attacks and defenses. In: 20th USENIX Conf. on Security, SEC 2011 (2011)
19. Fragkaki, E., Bauer, L., Jia, L., Swasey, D.: Modeling and Enhancing Android's Permission System. In: Foresti, S., Yung, M., Martinelli, F. (eds.) ESORICS 2012. LNCS, vol. 7459, pp. 1–18. Springer, Heidelberg (2012)
20. Fuchs, A., Chaudhuri, A., Foster, J.S.: SCanDroid: Automated security certification of android applications. Technical report, U. of Maryland College Park (2009)
21. Google. Android 4.1 Compatibility Definition. Android Compatibility Program, Rev. 2 (September 7, 2012)
22. Hornyack, P., Han, S., Jung, J., Schechter, S., Wetherall, D.: These aren't the droids you're looking for: retrofitting android to protect data from imperious applications. In: 18th ACM Conf. on Computer and Comm. Security, CCS 2011 (2011)
23. Nauman, M., Khan, S., Zhang, X.: Apex: extending android permission model and enforcement with user-defined runtime constraints. In: 5th ACM Symp. on Information, Computer and Communications Security, ASIACCS 2010 (2010)
24. NielsenWire. State of the appnation - a year of change and growth in U.S. smartphones (May 16, 2012), blog.nielsen.com/nielsenwire/online_mobile/state-of-the-appnation-%E2%80%93-a-year-of-change-and-growth-in-u-s-smartphones/
25. Schreckling, D., Posegga, J., Köstler, J., Schaff, M.: Kynoid: Real-Time Enforcement of Fine-Grained, User-Defined, and Data-Centric Security Policies for Android. In: Askoxylakis, I., Pöhls, H.C., Posegga, J. (eds.) WISTP 2012. LNCS, vol. 7322, pp. 208–223. Springer, Heidelberg (2012)
26. Shekhar, S., Dietz, M., Wallach, D.: Adsplit: separating smartphone advertising from applications. In: 21st USENIX Conf. on Security, SEC 2012 (2012)
27. Stevens, R., Gibler, C., Crussell, J., Erickson, J., Chen, H.: Investigating user privacy in android ad libraries. In: MoST 2012: Mobile Security Technologies (2012)
28. Varma, K.: Security permissions in android. Krishnaraj Varma's Blog (October 3, 2010), www.krvarma.com/2010/10/security-permissions-in-android/ (accessed October 9, 2012)
29. Xu, R., Saïdi, H., Anderson, R.: Aurasium: practical policy enforcement for android applications. In: 21st USENIX Conf. on Security, SEC 2012 (2012)

Model-Based Usage Control Policy Derivation

Prachi Kumari and Alexander Pretschner

Technische Universität München, Germany
{kumari,pretschn}@cs.tum.edu

Abstract. Usage control is concerned with how data is used after access
to it has been granted. In existing usage control enforcement frameworks,
policies are assumed to exist and the derivation of implementation-level
policies from specification-level policies has not been looked into. This
work fills this gap. One challenge in the derivation of policies is the
absence of clear semantics of high-level domain-specific constructs like
data and action. In this paper we present a model-based refinement of
these constructs. Using this refinement, we translate usage control policies from the specification to the implementation level. We also provide
methodological guidance to partially automate this translation.

1 Introduction

Usage control systems provide means to specify and enforce policies about the
future usage of data. Usage control requirements have been enforced for various
policy languages [1–8], at [9–18] and across [19] different layers of abstraction in
various types of systems [20,21]. The focus there has been on the implementation
of policy monitors. How policies are specified, translated or instantiated, has not
been addressed. The challenge is that system implementations of usage control
policies might not always adequately reflect end user requirements. This is due
to several reasons, one of which is the problem of mapping concepts in the end
user's domain to technical events and artifacts. For instance, the semantics of
basic operators such as "copy" or "delete", which are fundamental for specifying
usage control policies, tend to vary according to domain context and can be
mapped to different sets of system events. This might wrongly allow events
that should have been inhibited and block those that should have been allowed.
Thus, in the absence of clear semantics of actions in an application context, it
is impossible to define and enforce usage control requirements in a way that is
unambiguous. This is the problem that we address in this paper.

We present a model-based policy derivation that combines usage control enforcement with data and action refinement. Policies are supposed to be specified by **end users** and translated using technical details provided by a more
sophisticated user whom we call the **power user**. The translation process is
semi-automated because it requires intervention from power users at specific
points. One use case is from a web-based social network (WBSN) where an end
user Alice would like to exercise control over copies of her data by other users.
She would specify "do not copy my photos" in a user-friendly way. This policy

J. Jürjens, B. Livshits, and R. Scandariato (Eds.): ESSoS 2013, LNCS 7781, pp. 58–74, 2013.
© Springer-Verlag Berlin Heidelberg 2013

would then be translated, deployed and enforced at all client-side machines that access Alice's data. We show the step-by-step translation of this policy in the rest of this paper. It is organized along five steps:

Step 1: Specification of policies We start with an overview of a policy language [19] that is used to express constraints on the future usage of data ("don't copy photos," "delete document after 30 days," "play video at most 5 times," etc.) These requirements are called **specification-level policies**.

Step 2: Refinement of actions We express specification-level policies in terms of high-level actions like "delete" or "copy." For enforcement, we must refine these actions into their technical counterparts. Intuitively, the semantics of actions vary according to the domain context. Therefore any solution that caters to the semantics of actions must address the problem at the domain level. We recap a domain-specific meta-model from the literature [22] that distinguishes between abstract and concrete events and refine the former to the latter (no formal semantics have been given to the refinements in the foundational work).

Step 3: Semantics of action refinement We combine the usage control model and the domain meta-model to specify the formal semantics of action refinement.

Step 4: From specification-level policies to ECA rules **Implementation-level policies** are rules of event-condition-action (ECA) form that execute an *action* when a trigger *event* takes place and the respective *condition* evaluates to true. As real systems cannot look into the future, the condition part of the ECA rules must be expressed in past tense. We provide a methodological guidance for automated transformation of specification-level policies to ECA rules.

Step 5: Example translation We present the translation of our example policy "don't copy photos" for enforcement in multiple systems.

We have deliberately not considered the dynamic nature of systems in this paper; systems structures are assumed to be static. Though unrealistic, this assumption is reasonable to narrow the scope for initial results.

This work provides semantics to abstract constructs in end-user policies by modeling the basis for such semantics. It is not possible to check the correctness of the semantics that adhere to our meta-model if they indeed correspond to the idea in the end user's mind. Hence we do not discuss any theorems to check if the semantics given by the power user are indeed correct.

Problem. In sum, we tackle two problems in this paper. The first one is the fundamental problem of the lack of semantics of high-level actions in usage control policies. The second one concerns the problem of transforming specification-level policies to implementation-level policies in an automated manner.

Solution. We present a model-based translation schema for high-level actions, taking into account the different representations of data and the potential data flow through a concrete system.

Contribution. We are not aware of any work that provides a semi-automated translation of specification-level usage control policies into implementation-level policies in a generic, domain- and system-independent way.

Organization. In §2 we recap a usage control model and a domain meta-model from the literature which we combine in §3 for action refinement. §4 backs our work with a detailed example. §5 puts our work in context and §6 concludes.

2   Background

Step 1: Specification of usage control policies. End user policies are expressed in OSL (originally described in [6]), a policy specification language that combines classical propositional operators with future-time temporal and cardinality operators. To specify and enforce policies on abstract data, the original usage control model was extended to distinguish between *data* (photo, song etc.) and its technical representations called *containers* (files, windows, records etc.) [19]. Enforcement of policies on data is done through data flow tracking. Possible data flows are defined by a transition relation on system states; actual data flows are monitored on the grounds of this relation. Formally, we consider systems $(P, Data, Event, Container, \Sigma, \sigma_i, \varrho)$ where P is a set of principals, *Data* is a set of data elements, *Event* is the set of events, *Container* is a set of data containers, Σ is the set of states of the system with σ_i being the initial state, and ϱ is the state transition function. System states are defined by a tuple of three mappings between data, containers and container identifiers: a *storage function* of type $Container \rightarrow \mathbb{P}(Data)$ that reflects which container stores what data; an *alias function* of type $Container \rightarrow \mathbb{P}(Container)$ that captures the fact that some containers may implicitly get updated whenever other containers do; and a *naming function* that provides names for containers and that is of type $F \rightarrow Container$, where F is a set of identifiers. The system's state space is defined as $\Sigma = (Container \rightarrow \mathbb{P}(Data)) \times (Container \rightarrow \mathbb{P}(Container)) \times (F \rightarrow Container)$ with the initial state $\sigma_i = (\varnothing, \varnothing, \varnothing)$. $Trace = \mathbb{N} \rightarrow (\Sigma \times \mathbb{P}(Event))$ captures both events and the information state at a moment in time. Transitions between two states are given by $\varrho : \Sigma \times \mathbb{P}(Event) \rightarrow \Sigma$. At any given point of time, the state of the system is computed using a recursive function $states : (Trace \times \mathbb{N}) \rightarrow \Sigma$ which in turn is defined as $states(t, 0) = \sigma_i$ and $n > 0 \Rightarrow states(t, n) = \varrho(states(t, n - 1), t(n - 1))$.

Policies are expressed in terms of *parameterized* events on data and container. Each event belongs to the set $Event \subseteq EventName \times (ParamName \rightarrow ParamValue)$. Data and containers are parameter values, belonging to disjoint sets *Data* and *Container*. Events are classified as *dataUsage* when they apply to a data object (reserved parameter *obj*) and *containerUsage* if they apply to a container object. The specification language is Φ^+ (+ for future), distinguishing between purely propositional (Ψ) and temporal and cardinality operators:

$$\Psi ::= \underline{true} \,|\, \underline{false} \,|\, E(Event) \,|\, T(Event) \,|\, \underline{not}(\Psi) \,|\, \underline{and}(\Psi, \Psi) \,|\, \underline{or}(\Psi, \Psi) \,|\, \underline{implies}(\Psi, \Psi)$$
$$\Phi^+ ::= \Psi \,|\, \underline{not}(\Phi^+) \,|\, \underline{and}(\Phi^+, \Phi^+) \,|\, \underline{or}(\Phi^+, \Phi^+) \,|\, \underline{implies}(\Phi^+, \Phi^+) \,|$$
$$\underline{until}(\Phi^+, \Phi^+) \,|\, \underline{after}(\mathbb{N}, \Phi^+) \,|\, \underline{within}(\mathbb{N}, \Phi^+) \,|\, \underline{during}(\mathbb{N}, \Phi^+) \,|\, \underline{always}(\Phi^+) \,|$$
$$\underline{repmax}(\mathbb{N}, \Psi) \,|\, \underline{replim}(\mathbb{N}, \mathbb{N}, \mathbb{N}, \Psi) \,|\, \underline{repuntil}(\mathbb{N}, \Psi, \Phi^+)$$

As events might be modified or blocked for enforcement, distinction between *attempted/desired* and *actual* events is needed. Formulas of the form $E(\cdot)$ and

$T(\cdot)$ denote actual and desired events; _not, and, or, implies_ have their intuitive semantics; _until_ is the weak until from LTL; the _always_ operator is intuitive; _after_(n, a) is true if a becomes true after n time steps; _within_(n, a) is true if a holds true at least once in n timepsteps, whereas _during_(n, a) is true only when a is constantly true in n timesteps. _repmax_(n, a) specifies that a must be true at most n times in the future; _replim_(l, m, n, a) specifies lower(l) and upper(m) bounds on repetitions of a in n timesteps and _repuntil_(n, a, b) limits the maximal number of times a holds until b holds.

Sometimes, it is convenient to specify policies not in terms of events but in terms of states a system must or must not enter. E.g., our example policy in §1, "don't copy photos," would mean that in an operating system, all sequences of system calls corresponding to "copy" actions must be inhibited. But infinitely many such sequences can achieve the effect of "copy," and it is infeasible to come up with a complete list of all of them. Instead, the same requirement can be expressed as, "data must not leave a specific set of containers." To allow this type of policies, three operators Φ_i have been added to Φ^+ [9,19]:

$$\Phi_i ::= \underline{isNotIn}(Data, \mathbb{P}\ Container)\ |\ \underline{isCombinedWith}(Data, Data)\ |$$
$$\underline{isOnlyIn}(Data, \mathbb{P}\ Container)$$

where _isNotIn_$(Data, \mathbb{P}\ Container)$ is true if data is not in a specific set of containers; _isCombinedWith_$(Data, Data)$ is true if two data items are stored in the same container; and _isOnlyIn_$(Data, \mathbb{P}\ Container)$ is true if data is only in a specified set of containers. The extended language is $\Phi_i^+ = \Phi^+ \cup \Phi_i$ with semantics $\models_i^+ \subseteq (Trace \times \mathbb{N}) \times \Phi_i^+$.

ECA rules, that we need for system-level enforcement, are specified in the past temporal logic Φ^- with added state-based operators, Φ_i. The extended past-time OSL is $\Phi_i^- = \Phi^- \cup \Phi_i$ with semantics $\models_i^- \subseteq (Trace \times \mathbb{N}) \times \Phi_i^-$.

$$\Phi^- ::= \Psi\ |\ \underline{not}^-(\Phi^-)\ |\ \underline{and}^-(\Phi^-, \Phi^-)\ |\ \underline{or}^-(\Phi^-, \Phi^-)\ |\ \underline{implies}^-(\Phi^-, \Phi^-)\ |$$
$$\underline{since}^-(\Phi^-, \Phi^-)\ |\ \underline{before}^-(\mathbb{N}, \Phi^-)\ |\ \underline{within}^-(\mathbb{N}, \Phi^-)\ |\ \underline{during}^-(\mathbb{N}, \Phi^-)\ |$$
$$\underline{always}^-(\Phi^-)\ |\ \underline{repmax}^-(\mathbb{N}, \Psi)\ |\ \underline{replim}^-(\mathbb{N}, \mathbb{N}, \mathbb{N}, \Psi)\ |\ \underline{repsince}^-(\mathbb{N}, \Psi, \Phi^-)$$

since$^-(a, b)$ is true if b has been true ever since a happened; _before_$^-(n, a)$ is true if a was true n time steps ago; _within_$^-$, _during_$^-$ and _always_$^-$ are intuitive, given the semantics of their future-time duals. _repmax_$^-(n, a)$ specifies that a has been true at most n times in the past; _replim_$^-(l, m, n, a)$ specifies a lower(l) and an upper limit(m) upon repetitions of a in the last n timesteps; and _repsince_$^-(n, a, b)$ specifies that b has been true at most n times since a became true.

Step 2: Refinement of actions. In the usage control model of step 1, both abstract actions and their technical counterparts are called _events_. But to refine actions, we must be able to distinguish between action (copy, delete etc.) and its technical representations (generic copy file or delete file; or more specifically, read, write, unlink systems calls in unix). Our domain meta-model [22], reproduced in Fig. 1, distinguishes among user-intelligible high-level actions on data like "copy photo" at the platform-independent (PIM) layer; corresponding implementation-independent technical representations (called **transformers**) like "take screenshot" at the platform-specific (PSM) layer; and the specific implementations of these transformers like "getImage()" function in the X11

windowing system at the implementation-specific (ISM) layer. Mappings between various components at different layers in the model provide the semantics
of a high level action in terms of a number of mapped transformers. As an example, the meta-model is instantiated for the refinement of copy in WBSN domain
(Figure 2). The "copy photo" part of Alice's WBSN policy would be refined in
this model as "copy&paste DOM element" and "screenshot of window" at the
PSM layer; and at the ISM layer as "copy_cmd on HTML element" in Firefox web
browser and as "getImage" function on a drawable in X11 windowing system.

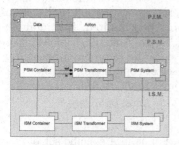

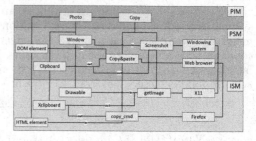

Fig. 1. The Domain Meta-model   **Fig. 2.** A WBSN instance of Fig 1

The concepts of data and action (PIM layer) and containers and transformers
(PSM and ISM layers) are also present in the usage control model, with some
differences. Firstly, the above usage control model uses the term *event* to refer to
both actions and transformers. In the domain meta-model, there is a clear distinction between the two. Secondly, in the domain meta-model, these constructs
have been grouped according to the level of technical detail they encompass.
Thus data and action form the PIM part whereas container and transformer
form the ISM part in the meta-model. Thirdly, to systematically reach from
elements of PIM to ISM, another layer of detail that maps the two, called the
PSM layer, is introduced in the meta-model. This is motivated by the systematic
translation requirement that the platform-specific result of a transformer on a
container must remain the same, irrespective of the implementations. For example, deleting a file can be achieved in many ways. But by defining it as "overwrite
file with random bytes OR remove file" at the PSM level narrows down the interpretation of deleting a file irrespective of the file system implementations.

The presentation of the domain meta-model [22] also contains a model-based,
semi-automated approach to policy translation. However, that work discusses
initial results at a high level and does not explain the exact relationship between
actions and transformers. Actions are mapped to transformers using UML associations, but no semantics has been given to these associations. So we do not
know what high-level actions like copy mean in terms of transformers. For example, looking at the system calls executed in a Unix operating system, we cannot
know if a copy action has indeed taken place because we do not know if copy
corresponds to the set or the sequence of these system calls. Even if we know

that it is the sequence, we do not know how the sequence is to be interpreted: if the system calls must happen one after the other or if some other executions can take place between any two of them. Reference [22] also does not relate high-level actions to system states: the authors only mention a refinement of the former in terms of the latter for cases where transformer-based refinements are not sufficient; they do not explain how this refinement is achieved (step 3b below). To address these issues, we combine this domain meta-model with the usage control model of step 1 to use the concept of system states and the semantics of language in terms of traces to formally refine actions and translate specification-level policies.

3 A Combined Model

Data is the set of all data, *Action* is the set of all actions, *PSMContainer* is the set of all PSM containers, *PSMTransformer* is the set of all PSM transformers and so on. *Event* is the set of all actions and transformers at all levels in the domain model: $Event = (Action \cup PSMTransformer \cup ISMTransformer)$. Associations between the elements of these sets are functions. So, $dataPotentiallyIn : Data \rightarrow \mathbb{P}(PSMContainer)$ maps data to a set of PSM containers that potentially store that data and $containerImplementedAs : PSMContainer \rightarrow \mathbb{P}(ISMContainer)$ gives a further refinement of PSM container in terms of a set of ISM containers that actually store data. Additionally, transformers are functions that modify respective containers:

$PSMTransformer : \mathbb{P}\,PSMContainer \rightarrow \mathbb{P}\,PSMContainer$
$ISMTransformer : \mathbb{P}\,ISMContainer \rightarrow \mathbb{P}\,ISMContainer$

Function $inputContainer : PSMTransformer \nrightarrow \mathbb{P}\,PSMContainer$ gives all containers modified by a PSM transformer. $inputContainer$ is overloaded to get input containers of ISM transformers. While the refinement of data is straightforward, actions can be refined in two ways: *SETrefmnt* maps an action to a set of PSM transformers with the intuitive semantics that any one of the mapped transformers corresponds to the high-level action; *SEQrefmnt* maps an action to a sequence of PSM transformers: all of the specified transformers in the particular sequence correspond to the high-level action. As PSM and ISM transformers can be further refined, both within and across their respective levels in the meta-model, their refinement functions are overloaded to express both of these refinements. *SETrefmnt* and *SEQrefmnt* express intra-level refinements

$SETrefmnt : PSMTransformer \rightarrow \mathbb{P}\,PSMTransformer$
$SETrefmnt : ISMTransformer \rightarrow \mathbb{P}\,ISMTransformer$
$SEQrefmnt : PSMTransformer \rightarrow seq\,PSMTransformer$
$SEQrefmnt : ISMTransformer \rightarrow seq\,ISMTransformer$

and, *crossSETrefmnt* and *crossSEQrefmnt* express inter-level refinements.

$crossSETrefmnt : Action \rightarrow \mathbb{P}\,PSMTransformer$
$crossSETrefmnt : PSMTransformer \rightarrow \mathbb{P}\,ISMTransformer$
$crossSEQrefmnt : Action \rightarrow seq\,PSMTransformer$
$crossSEQrefmnt : PSMTransformer \rightarrow seq\,ISMTransformer$

In a specific domain model, the PSM level talks about the static, design-time system while the ISM level talks about the concrete system at the runtime.

Data Storage. The storage function in the usage control model tells which container stores what data. For the translation of actions on specific data, we need the reverse relationship; we need to know where a specific data is stored in a particular moment in time. Function $dataActuallyIn : Data \times \Sigma \to \mathbb{P}\ ISMContainer$ gives this information:

$$\forall\, d \in Data;\ t \in Trace;\ n \in \mathbb{N};\ \sigma \in \Sigma \bullet \sigma = states(t, n) \wedge$$
$$dataActuallyIn(d, \sigma) = \{c \in ISMContainer \mid d \in \sigma.1(c)\}$$

where $\sigma.1$ denotes the projection on the first component of σ.

Remember that formulas of the form $E(\cdot)$ and $T(\cdot)$ denote actual and desired events in the OSL. Therefore, refinement of actions corresponds to the translation of OSL formulas of the form $E(\cdot)$ and $T(\cdot)$. A high-level action is refined in two ways: **firstly**, in terms of sets/sequences of transformers using function τ_{ev}; **secondly**, in terms of system states using function τ_{state}. We combine both refinements to get the **complete** refinement of a high-level action.

Step 3a: Action Refinement using Transformers. As it is impossible to predict the length of executions between any two members of a sequence of transformers in real systems, we allow arbitrary executions between any two members of a sequence of applicable transformers in *SEQrefmnt*. This introduces liveness in our action refinement definitions. Because indefinite past can be checked in a running system as opposed to indefinite future, we first translate a specification-level policy from future to past tense and then execute action refinement. For this reason, our action refinement functions act on, and are formalized, using past-time OSL operators. A translation function $\tau_p : \Phi^+ \to \Phi^-$ that works along the lines of the methodological guidance provided in [22], translates a formula in Φ^+ to another in Φ^-. To express indefinite past, we use $\underline{eventually}^-$, semantically equivalent to $\underline{not}^-(\underline{always}^-(\underline{not}^-))$ in the language Φ_i^-. Intuitively, $\underline{eventually}^-(\varphi)$ is true if the formula φ was true at least once in the past.

$\tau_{ev} : \Phi^- \times \Sigma \to \Phi^-$ translates an action into sets/sequences of transformers that are further refined using $\pi_{ev} : \Phi^- \times \Sigma \to \Phi^-$. The system state ($\Sigma$) provides the knowledge of data storage in specific containers and is filled in by the "higher" translation function τ_{action}, defined later in this paper.

We refine high-level actions by taking into account all representations of data in a concrete system. Therefore, only those transformers that modify containers where data *may* reside, refine the corresponding high-level action. If the data on which an action operates, cannot be stored in a particular container, all transformers that operate on this container are left out of the refinement process. For example, a copy action is refined into the set {copyFile(file), takeScreenshot(window)} at the PSM level. When the data is a song and the policy addresses copy(song), the action is refined into set {copyFile(file)} rather than {copyFile(file), takeScreenshot(window)}. This is because the other transformer operates on windows where a song cannot be stored. In this example, {copyFile(file)} is the set of **applicable transformers**. The set of applicable transformers for a data $(appTransformer : Data \nrightarrow \mathbb{P}\, PSMTransformer)$ is computed as follows:

$$\forall\, d \in Data \bullet appTransformer(d) =$$
$$\{t \in PSMTransformer \mid inputContainer(t) \subseteq dataPotentiallyIn(d)\}$$

Using set and sequence mappings from action to PSM transformers that modify potential storage of data object of the action (via $\mathsf{ran}\ SEQrefmnt(e) \cap appTransformer(d)$ etc.), we compute action refinement upto the lowest level in the platform-specific model (ran is the standard operation for sequences [23] that gives the set of objects which are elements of the sequence)

$$\forall\, s \in Trace;\ x \in \mathbb{N};\ \sigma \in \Sigma \bullet \sigma = states(s, x) \Rightarrow$$
$$\forall\, d \in Data;\ e \in Event;\ \{t_1, .., t_n\} \in \mathbb{P}(PSMTransformer);\ \varphi \in \Phi^- \bullet$$
$$\tau_{ev}(E(e, \{(obj, d)\}), \sigma) = \varphi \Leftrightarrow$$
$$\varphi = \underline{and}^-(\tau_{ev}(E(t_n, \{(obj, d)\}), \sigma), \underline{eventually}^-(\underline{and}^-(\tau_{ev}(E(t_{n-1}, \{(obj, d)\}), \sigma),$$
$$...\underline{eventually}^-(\tau_{ev}(E(t_1, \{(obj, d)\}), \sigma))...)))\ \wedge$$
$$(\{t_1, .., t_n\} = \mathsf{ran}\ SEQrefmnt(e) \cap appTransformer(d)\ \vee$$
$$\{t_1, .., t_n\} = \mathsf{ran}\ crossSEQrefmnt(e) \cap appTransformer(d))$$
$$\vee\ \varphi = \underline{or}^-(\tau_{ev}(E(t_1, \{(obj, d)\}), \sigma), \underline{or}^-(\tau_{ev}(E(t_2, \{(obj, d)\}), \sigma),$$
$$..., \tau_{ev}(E(t_n, \{(obj, d)\}), \sigma)))\ \wedge$$
$$(\{t_1, ..., t_n\} = SETrefmnt(e) \cap appTransformer(d)\ \vee$$
$$\{t_1, ..., t_n\} = crossSETrefmnt(e) \cap appTransformer(d))$$
$$\vee\ \varphi = \pi_{ev}(E(e, \{(obj, d)\}), \sigma)$$

π_{ev} further refines these transformers till the ISM level. Meanings of sequence and set refinement remain the same. From all the possible input containers, mapped transformers act on only those containers that indeed store the specific data object ($c \in inputContainer(t) \cap dataActuallyIn(d, \sigma)$):

$$\forall\, s \in Trace;\ x \in \mathbb{N};\ \sigma \in \Sigma \bullet \sigma = states(s, x) \wedge$$
$$\forall\, d \in Data;\ e \in Event;\ \{t_1, .., t_n\} \in \mathbb{P}(ISMTransformer);\ t \in \{t_1, .., t_n\};$$
$$c \in (inputContainer(t) \cap dataActuallyIn(d, \sigma));\ \varphi \in \Phi^- \bullet$$
$$\pi_{ev}(E(e, \{(obj, d)\}), \sigma) = \varphi \Leftrightarrow$$
$$\varphi = \underline{and}^-(E(t_n, \{(obj, c)\}), \underline{eventually}^-(\underline{and}^-(E(t_{n-1}, \{(obj, c)\}),$$
$$...\underline{eventually}^-(E(t_1, \{(obj, c)\}))...)))\ \wedge \langle t_1, .., t_n \rangle = crossSEQrefmnt(e)$$
$$\vee\ \varphi = \underline{or}^-(E(t_1, \{(obj, c)\}), \underline{or}^-(E(t_2, \{(obj, c)\}), ..., E(t_n, \{(obj, c)\})))\ \wedge$$
$$\{t_1, ..., t_n\} = crossSETrefmnt(e)$$
$$\vee\ \varphi = false$$

Step 3b: Action Refinement Using State. We have seen in §2 that expressing the semantics of high level actions in terms of sets/sequences of transformers might not be the best approach in many cases because one high-level action can be refined to infinitely many sequences of transformers at the system level. To address this issue, we define another refinement of action, τ_{state}, using state-based operators of Φ_i. This translation captures the state a system reaches when a high-level action is executed on some data. The power user models the execution of each action and defines the resultant state as a *StateFormula* for each high-level action. Intuitively, when no resultant state is defined for an action, its state-based translation is *false*.

To define the set of all possible events that can occur in a concrete system, event declarations were introduced in [6]. These event declarations are purely syntactic and are given by $EventDecl == EventName \times EventClass \times (ParamName \nrightarrow \mathbb{P}\, ParamValue)$. To express the state-based refinement of an action, we modify this event declaration by adding $StateFormula$ to it. Thus, an event declaration is given by the event name, the event class, a partial function that defines the name and possible values of each possible parameter and, the resultant state formula that gives the state-based refinement of the event.

$EventDecl ==$

$EventName \times EventClass \times (ParamName \nrightarrow \mathbb{P}\, ParamValue) \times StateFormula$

The relationship between an action and its declaration is **bijective**. For the state-based translation of action, a function $getStateFormula$ fetches the resultant state from the declaration of the specific action:

$getStateFormula : Action \nrightarrow StateFormula$

$\forall\, a \in Action;\ ed \in EventDecl \bullet getStateFormula(a) = ed.4 \Leftrightarrow a.1 = ed.1$

The resultant state formula for an action is statically defined, before the action is used to specify policies. Actual data objects and their containers are known only when a policy is deployed in a concrete system. So resultant state formulas must address all potential data and their containers. To specify potential data in the state-based formula, we use variables which are substituted by actual data when a policy is specified. To specify containers that potentially store data, we extend the language Φ_i to include state-based operators on PSM containers. At runtime, respective ISM containers are extracted via function $containerImplementedAs$, introduced in the beginning of §3.

PSM Containers in state-based operators. To specify PSM containers in state-based operators, we classify ISM containers according to the PSM containers they implement. So each ISM container belongs to a *container class* that is a PSM container. Function $getContainerClass : ISMContainer \rightarrow PSMContainer$ extracts the class of a container using function $containerImplementedAs$:

$\forall\, c \in ISMContainer;\ cl \in PSMContainer \bullet$

$getContainerClass(c) = cl \Leftrightarrow c \in containerImplementedAs(cl)$

We extend the language by overloading two operators with PSM containers: $isNotIn(Data, \mathbb{P}\, PSMContainer)$ and $isOnlyIn(Data, \mathbb{P}\, PSMContainer)$. Intuitively, $isNotIn(d, Cl)$ is true if data d is not in any container whose class is in set Cl. This operator is useful for defining state-based refinement of actions like copy or print. For example, if print is refined as $not(isNotIn(d, \{print_{cont}\}))$ where $print_{cont}$ represents the class of printer containers. When data d flows into any container that belongs to this class, the enforcement infrastructure would recognize a print. Similarly, $isOnlyIn(d, Cl)$ is true when data is restricted to specific classes of containers. This is useful to express semantics of *weak deletion* where data is not actually deleted but only quarantined. We did not find any use case where the semantics of $isCombinedWith(d, d)$ need to be specified using container classes. The new operators are added to the language Φ_i:

$\Phi_i ::= isNotIn(Data, \mathbb{P}\, ISMContainer) \mid isNotIn(Data, \mathbb{P}\, PSMContainer) \mid$
$\quad isOnlyIn(Data, \mathbb{P}\, ISMContainer) \mid isOnlyIn(Data, \mathbb{P}\, PSMContainer) \mid$
$\quad isCombinedWith(Data, Data)$

The semantics of Φ_i is $\models_i \subseteq (Trace \times \mathbb{N}) \times \Phi_i$, shown in Figure 3.

Variables in OSL. Resultant state formulas are expressed in terms of potential data and their containers because actual data and their containers are known only when a policy is deployed in a concrete system. To specify state formulas with potential data items, we introduce variables in the language. Only variable data is needed; potential containers are specified using PSM containers. In case of *isCombinedWith*, the first data is variable, the second data is given by the power user. The respective language is Φ_{iv}.

$Var ::= V(\mathbb{N}_1)$
$VarData == Var \cup Data$
$\Phi_{iv} ::= isNotIn(VarData, \mathbb{P}\, ISMContainer) \mid isNotIn(VarData, \mathbb{P}\, PSMContainer) \mid$
$\quad isOnlyIn(VarData, \mathbb{P}\, ISMContainer) \mid isOnlyIn(VarData, \mathbb{P}\, PSMContainer) \mid$
$\quad isCombinedWith(VarData, Data)$

Elements from Φ_{iv} are instantiated into elements from Φ_i using **substitution**.

Finally, the refinement of actions in terms of states is achieved using τ_{state} : $\Phi^- \to \Phi_i^-$. When an action is refined, the variable in the respective state formula is substituted by the value of the *obj* parameter of the action.

$\forall a \in Action;\ d \in Data;\ \varphi \in \Phi_{iv};\ vd \in VarData \bullet$

$$\tau_{state}(E(a, \{(obj, d)\})) = \begin{cases} \varphi[d/vd] & \text{if}(\varphi = getStateFormula(a)) \\ false & \text{otherwise} \end{cases}$$

We have defined the refinement of a high-level action in terms of sets/sequences of transformers (using function τ_{ev}) and in terms of system states (using function τ_{state}). We now combine both functions to express the "complete" refinement of a high-level action, given by $\tau_{action} : (\Phi^- \times \Sigma) \to \Phi_i^-$. Intuitively, at least one of the refinements is needed to express a high-level action in a concrete system. Hence the disjunction ($\underline{or}^-$) over the refinements (Figure 4).

Step 4: From specification-level policies to enforcement mechanisms
Policy specification and translation is semi-automated with two roles of users: the

---

$\forall t \in Trace;\ n \in \mathbb{N};\ \varphi \in \Phi_i;\ \sigma \in \Sigma \bullet\ (t, n) \models_i \varphi \Leftrightarrow \sigma = states(t, n) \wedge$
$\quad \exists d \in Data,\ C \in \mathbb{P}\, ISMContainer \bullet \varphi = \underline{isNotIn}(d, C) \wedge$
$\qquad \forall c' \in ISMContainer \bullet d \in \sigma.1(c') \Rightarrow (c' \notin C)$
$\quad \exists d \in Data,\ Cl \in \mathbb{P}\, PSMContainer \bullet \varphi = \underline{isNotIn}(d, Cl) \wedge$
$\qquad \forall c' \in ISMContainer \bullet d \in \sigma.1(c') \Rightarrow (getContainerClass(c') \notin Cl)$
$\quad \vee \exists d \in Data,\ C \in \mathbb{P}\, ISMContainer \bullet \varphi = \underline{isOnlyIn}(d, C) \wedge$
$\qquad \forall c' \in ISMContainer \bullet d \in \sigma.1(c') \Rightarrow (c' \in C)$
$\quad \vee \exists d \in Data,\ Cl \in \mathbb{P}\, PSMContainer \bullet \varphi = \underline{isOnlyIn}(d, Cl) \wedge$
$\qquad \forall c' \in ISMContainer \bullet d \in \sigma.1(c') \Rightarrow (getContainerClass(c') \in Cl)$
$\quad \vee \exists d_1, d_2 \in Data \bullet \varphi = \underline{isCombinedWith}(d_1, d_2) \wedge$
$\qquad \exists c' \in ISMContainer \bullet d_1 \in \sigma.1(c') \wedge d_2 \in \sigma.1(c')$

Fig. 3. Semantics of Φ_i

$\forall\, t \in Trace;\ n \in \mathbb{N};\ \sigma \in \Sigma \bullet \sigma = states(t, n)\ \wedge$

$\forall\, d \in Data;\ a \in Action;\ \psi \in \Phi^{-};\ \varphi \in \Phi_{i}^{-} \bullet \tau_{action}(\psi, \sigma) = \varphi \Leftrightarrow$

$\psi \in \{true, false\} \wedge (\varphi = \psi)$

$\vee\ \psi = E(a) \wedge (\varphi = \underline{or}^{-}(\tau_{state}(E(a)), \tau_{ev}(E(a), \sigma)))$

$\vee\ \psi = T(a) \wedge (\varphi = \underline{or}^{-}(\tau_{state}(T(a)), \tau_{ev}(T(a), \sigma)))$

$\vee\ \exists \chi \in \Phi^{-} \bullet \psi \in \{\underline{not}(\chi), \underline{not}^{-}(\chi)\} \wedge (\varphi = \underline{not}^{-}(\tau_{action}(\chi, \sigma)))$

$\vee\ \exists \chi, \xi \in \Phi^{-} \bullet \psi \in \{\underline{or}(\chi, \xi), \underline{or}^{-}(\chi, \xi)\} \wedge (\varphi = \underline{or}^{-}(\tau_{action}(\chi, \sigma), \tau_{action}(\xi, \sigma)))$

$\vee\ \exists \chi, \xi \in \Phi^{-} \bullet \psi \in \{\underline{and}(\chi, \xi), \underline{and}^{-}(\chi, \xi)\} \wedge (\varphi = \underline{and}^{-}(\tau_{action}(\chi, \sigma), \tau_{action}(\xi, \sigma)))$

$\vee\ \exists \chi, \xi \in \Phi^{-} \bullet \psi \in \{implies(\chi, \xi), implies^{-}(\chi, \xi)\} \wedge (\varphi = implies^{-}(\tau_{action}(\chi, \sigma), \tau_{action}(\xi, \sigma)))$

$\vee\ \exists \chi, \xi \in \Phi^{-} \bullet \psi = \underline{since}^{-}(\chi, \xi) \wedge (\varphi = \underline{since}^{-}(\tau_{action}(\chi, \sigma), \tau_{action}(\xi, \sigma)))$

$\vee\ \exists i \in \mathbb{N};\ \chi \in \Phi^{-} \bullet \psi = \underline{before}^{-}(i, \chi) \wedge (\varphi = \underline{before}^{-}(i, \tau_{action}(\chi, \sigma)))$

$\vee\ \exists \chi \in \Phi^{-} \bullet \psi = \underline{always}^{-}(\chi) \wedge (\varphi = \underline{always}^{-}(\tau_{action}(\chi, \sigma)))$

$\vee\ \exists i \in \mathbb{N};\ \chi \in \Phi^{-} \bullet \psi = \underline{within}^{-}(i, \chi) \wedge (\varphi = \underline{within}^{-}(i, \tau_{action}(\chi, \sigma)))$

$\vee\ \exists i \in \mathbb{N};\ \chi \in \Phi^{-} \bullet \psi = \underline{during}^{-}(i, \chi) \wedge (\varphi = \underline{during}^{-}(i, \tau_{action}(\chi, \sigma)))$

$\vee\ \exists i \in \mathbb{N};\ \chi \in \Phi^{-} \bullet \psi = \underline{repmax}^{-}(i, \chi) \wedge (\varphi = \underline{repmax}^{-}(i, \tau_{action}(\chi, \sigma)))$

$\vee\ \exists l, x, y \in \mathbb{N};\ \chi \in \Phi^{-} \bullet \psi = \underline{replim}^{-}(l, x, y, \chi) \wedge (\varphi = \underline{replim}^{-}(l, x, y, \tau_{action}(\chi, \sigma)))$

$\vee\ \exists i \in \mathbb{N};\ \chi, \xi \in \Phi^{-} \bullet \psi = \underline{repsince}^{-}(i, \chi, \xi) \wedge (\varphi = \underline{repsince}^{-}(i, \tau_{action}(\chi, \sigma), \tau_{action}(\xi, \sigma)))$

Fig. 4. Definition of τ_{action}

end user specifies usage control policies with constructs and templates defined by the more sophisticated power user §1.

In the **first step**, τ_p translates a future-time formula into another past-time formula [22]. In the **second step**, action refinement takes place. After action refinement, we get a complex, nested formula that is broken down to subformulas (Fischer Ladner closure) in the **third step** and each subformula is then mapped to the condition part of one ECA rule in the **fourth step**. Thus we get a set of ECA rules corresponding to one specification-level policy. One high-level policy can be enforced in many ways (allow/modify/inhibit/delay). For example, Alice's policy "don't copy photo" can be enforced by inhibiting every copy event; it can be enforced by modifying the original photo with one that shows an error message; it can also be enforced by delaying the event until a permission for copying has been granted by Alice. For this reason, the action part of ECA rules cannot be specified automatically. The generic format of ECA rules at the end of step 4 is as follows (where c is one subformula)

```
Event: any
Condition: c
Action: ALLOW/MODIFY/INHIBIT/DELAY
```

Intuitively, (later configured) action takes place when the corresponding condition c is true, irrespective of the trigger event. To limit the set of trigger events for each rule, whenever c is of the form $\underline{and}^{-}(E(e), x)$ or $\underline{and}^{-}(T(e), x)$ where x is an OSL formula, we move e to the trigger event part and only x is checked in the condition part of the ECA rule.

```
Event: e
Condition: x
Action: ALLOW/MODIFY/INHIBIT/DELAY
```

All the steps described above are automated. In the **fifth step**, the power user manually specifies the enforcement mechanism. We now describe in detail the translation of the example policy introduced in §1.

4   Example Translation

Step 5: The partial domain model with transformer-based refinement of "copy photo" is shown Figure 5. The distinction between set and sequence refinements of events is shown via links with arrowheads representing SETrefmnts and links with AND(S) gate -head representing SEQrefmnts. For state-based action

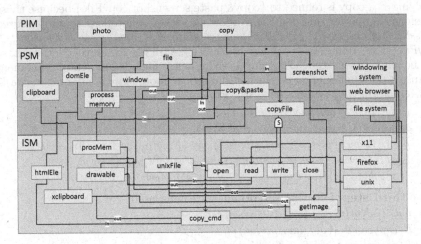

Fig. 5. Example domain model

refinement, state formula is defined in the event declaration. *Copy* in this context means data flows in clipboard containers; hence the respective state formula is $\underline{not}^{-}(\underline{isNotIn}(x, clipboard))$ where x is variable data and *clipboard* is the class of clipboard containers.

In our implementation, "don't copy photos" is specified by Alice in a block editor that uses the Open Blocks Java library [24]. When the data is sent to another user Bob, the respective policy is delivered to the policy translation point (PTP) which immediately translates and deploys the policy.

In the runtime, when policies are deployed, only concrete containers exist. So in our implementation, data is identified by the initial container in which it appears in the concrete system. With our usage control infrastructure, it is possible to track multiple representations of the same data at and across different abstract layers in a system. Hence, the PTP knows that Alice's photo is received by Bob at the web browser level in the initial container "img_profile" in Firefox; is stored in "myphoto.jpg" in the cache folder and rendered in window "0x1a00005" in X11.

The policy is $\underline{always}(\underline{not}(E(copy, \{(obj, img\_profile)\})))$ in OSL. It is of the form $\underline{always}(\varphi)$ where $\varphi = \underline{not}(E(copy, \{(obj, img\_profile)\}))$. τ_p gives us the past-time condition to be checked in the respective ECA rules. $\tau_p(\underline{always}(\varphi)) = \underline{and}^{-}(\underline{before}^{-}(1, \tau_p(\varphi)\underline{since}^{-}START), \underline{not}^{-}(\tau_p(\varphi)))$ where $START$ denotes the policy activation event [22]. This means that the respective ECA rule is triggered when φ has always been true since the policy was activated, except the current time-step. As $\varphi = \underline{not}(E(copy, \{(obj, img\_profile)\}))$, the ECA rule is triggered when $E(copy, \{(obj, img\_profile)\})$ is true.

The next step is action refinement as described in §3. $\tau_{action}(E(copy, \{(obj, img\_profile)\}), \sigma)$, where σ is the current state, works as follows: State-based refinement is achieved by substituting variable x with img_profile in the state formula:

$\tau_{state}(E(copy, \{(obj, img\_profile)\})) = \underline{not}^-(isNotIn(img\_profile, clipboard))$

Applying τ_{ev}, copy is refined to {copy&paste,screenshot,copyFile} because these transformers operate on {domEle,window,file} where photo is potentially stored $(crossSETrefmnt(copy) \cap appTransformer(photo) = \{copy\&paste, screenshot, copyFile\})$; π_{ev} refines each of these transformers till the ISM level. Note that, of the sequence $\langle open, read, write, close \rangle$, write is not included in action refinement because it does not operate on the file that stores photo $(inputContainer(write) \cap dataActuallyIn(myphoto.jpg, \sigma) = \varnothing)$. Finally,

$\tau_{action}(E(copy, \{(obj, img\_profile)\}), \sigma)$
$= \underline{or}^-(\underline{not}^-(isNotIn(img\_profile, clipboard)),$
$\quad \underline{or}^-(E(copy\_cmd, \{(obj, img\_profile)\}), \underline{or}^-(E(getImage, \{(obj, 0x1a00005)\}),$
$\quad (\underline{and}^-(E(close, \{(obj, myphoto.jpg)\}), \underline{eventually}^-(\underline{and}^-(E(read, \{(obj,$
$\quad\quad myphoto.jpg)\}), \underline{eventually}^-(E(open, \{(obj, myphoto.jpg)\}))))))))))$

In the **third step**, following subformulas are computed:
$\varphi_1 = \underline{not}^-(isNotIn(img\_profile, clipboard))$
$\varphi_2 = E(copy\_cmd, \{(obj, img\_profile)\})$
$\varphi_3 = E(getImage, \{(obj, 0x1a00005)\})$
$\varphi_4 = \underline{and}^-(E(close, \{(obj, myphoto.jpg)\}), \underline{eventually}^-(\underline{and}^-(E(read, \{(obj,$
$\quad\quad myphoto.jpg)\}), \underline{eventually}^-(E(open, \{(obj, myphoto.jpg)\})))))$

In the **fourth step**, generic ECA rules, as described above, are generated for each subformula. φ_2 and φ_3 are of the form $\underline{and}^-(E(e), true)$. So respective e becomes the trigger event as described above, and the condition part of the respective ECA rules is $true$. The specific action to be taken in each ECA rule is manually specified in the **fifth step**.

5   Related Work

The goal of this work is to automate the refinement of policies in the context of usage control. Policy refinement has been the focus of research since quite some time [25] and in the recent years, there have been various attempts towards automating it. Solutions have been based on refining policies using resource hierarchies [26], commitment (obligations) analysis [27], goal decomposition [28], data classification [29] and also from different perspectives viz. conflict prevention, where the focus has more been on the translation of constraints [30]. In [31] and [32], ontology-based refinement techniques are described for semi-automated translation of access control policies. In our work, such ontologies could be used at each level of the meta-model. In [33], authors have proposed a resource hierarchy meta-model for translating domain-specific elements in XACML policies for virtual organizations to generate corresponding resource-level policies. This is similar to our work in terms of the approach. However, the policies are refined from the abstract level (users, resources and applications) to the logical level (user ids, resource addresses and computational commands like read/write); further technical representations of policy elements in concrete systems are not considered. Another work which is quite similar to ours in terms of approach is described in [34]. This paper focuses on action decomposition in a policy refinement framework. Subjects perform operations on targets (services and devices) which are specified at a high level. Using a system model and a set of refinement rules, actions are decomposed and one higher level policy is refined into multiple policies. However, all elements (both abstract and concrete) of the system model are at the same level; which makes this approach similar to the ontology-based refinement. Also, in the last stage of refinement, policies are transformed into ECA rules. How this transformation is achieved is however not specified.

In almost all of the work on policy refinement, there has been some kind of distinction between the abstract entities at high level and the corresponding technical entities at lower levels. This approach of capturing details of a system with several levels of abstraction has been addressed in many architecture frameworks [35–37] and is also common in the embedded systems domain [38–40]. We have adopted a model-based approach which is analogous to the MDA viewpoints [41] with varying level of details at the computation-independent, platform-independent and platform-specific levels. A minor difference with the MDA approach is in the naming of the different layers. We have combined this approach with usage control concepts to refine policies.

The contribution w.r.t. to reference [22] is detailed out in §2, step 2.

6   Conclusion and Future Work

This paper describes a model-based policy refinement for usage control enforcement. Through this work, we have addressed the fundamental problem of the lack of semantics of actions like copy or delete. Additionally, we have provided a methodological guidance for transforming specification-level policies into implementation-level policies that configure enforcement mechanisms at different layers of abstraction. This helps translate policies in an automated manner.

For precise semantics of action refinement, we have combined an existing domain meta-model with a usage control model from the literature. The combined model captures both the static (all possible cases) and dynamic (one particular case with runtime information) aspects of concrete systems. The refinement of actions in this combined model is twofold: actions are refined to sets/sequences of low-level transformers and also to state-based formulas that describe the storage of data in containers. Refinement of actions is used to give semantics to specification-level policies in terms of a set of system traces. We have also provided methodological guidance to automate the policy translation: when future-time policies are translated to their past-time equivalents, the complex formula with all action refinements is decomposed into subformulas and mapped to the condition part of ECA rules.

It is hard to establish a notion of correctness between the semantics of low-level and high-level policies because the semantics of high-level propositions is not precisely defined but rather exists in the (end) user's mind. In fact, we see our translation procedure as a way to *define* the semantics of high-level policies by assigning machine-level events and state changes to high-level actions.

We have deliberately introduced a limitation in this paper: we have not considered the dynamic nature of systems. Adaptive policy translation is a topic of ongoing work. Another topic of current investigation is the evolution of policies [42] since we have not considered the fact that specification-level policies may also change from one receiver to another in a distributed setup.

References

1. Iannella, R. (ed.): Open Digital Rights Language v1.1 (2008), http://odrl.net/1.1/ODRL-11.pdf
2. Multimedia framework (MPEG-21) – Part 5: Rights Expression Language. ISO/IEC standard 21000-5:2004 (2004)
3. Ashley, P., Hada, S., Karjoth, G., Powers, C., Schunter, M.: Enterprise Privacy Authorization Language (EPAL 1.2). IBM Technical Report (2003)
4. Open Mobile Alliance. DRM Rights Expression Language V2.1 (2008), http://www.openmobilealliance.org/Technical/release_program/drm_v2_1.aspx
5. Zhang, X., Park, J., Parisi-Presicce, F., Sandhu, R.: A logical specification for usage control. In: Proc. SACMAT, pp. 1–10 (2004)
6. Hilty, M., Pretschner, A., Basin, D., Schaefer, C., Walter, T.: A Policy Language for Distributed Usage Control. In: Biskup, J., López, J. (eds.) ESORICS 2007. LNCS, vol. 4734, pp. 531–546. Springer, Heidelberg (2007)
7. Damianou, N., Dulay, N., Lupu, E., Sloman, M.: The Ponder Policy Specification Language. In: Sloman, M., Lobo, J., Lupu, E.C. (eds.) POLICY 2001. LNCS, vol. 1995, pp. 18–38. Springer, Heidelberg (2001)
8. W3C. The Platform for Privacy Preferences 1.1 (P3P1.1) Specification (2005), http://www.w3.org/TR/2005/WD-P3P11-20050104/
9. Harvan, M., Pretschner, A.: State-based Usage Control Enforcement with Data Flow Tracking using System Call Interposition. In: Proc. 3rd Intl. Conf. on Network and System Security, pp. 373–380 (2009)
10. Pretschner, A., Buechler, M., Harvan, M., Schaefer, C., Walter, T.: Usage control enforcement with data flow tracking for x11. In: Proc. STM 2009, pp. 124–137 (2009)
11. Dam, M., Jacobs, B., Lundblad, A., Piessens, F.: Security Monitor Inlining for Multithreaded Java. In: Drossopoulou, S. (ed.) ECOOP 2009. LNCS, vol. 5653, pp. 546–569. Springer, Heidelberg (2009)
12. Ion, I., Dragovic, B., Crispo, B.: Extending the Java Virtual Machine to Enforce Fine-Grained Security Policies in Mobile Devices. In: Proc. Annual Computer Security Applications Conference, pp. 233–242. IEEE Computer Society (2007)
13. Desmet, L., Joosen, W., Massacci, F., Naliuka, K., Philippaerts, P., Piessens, F., Vanoverberghe, D.: The S3MS.NET Run Time Monitor: Tool Demonstration. ENTCS 253(5), 153–159 (2009)
14. Erlingsson, U., Schneider, F.: SASI enforcement of security policies: A retrospective. In: Proc. New Security Paradigms Workshop, pp. 87–95 (1999)

15. Yee, B., Sehr, D., Dardyk, G., Chen, J., Muth, R., Ormandy, T., Okasaka, S., Narula, N., Fullagar, N.: Native Client: A Sandbox for Portable, Untrusted x86 Native Code. In: Proc. IEEE Symposium on Security and Privacy, pp. 79–93 (2009)
16. Gheorghe, G., Neuhaus, S., Crispo, B.: xESB: An Enterprise Service Bus for Access and Usage Control Policy Enforcement. In: Proc. ICTM (2010)
17. Egele, M., Kruegel, C., Kirda, E., Yin, H., Song, D.: Dynamic spyware analysis. In: Proceedings of USENIX Annual Technical Conference (June 2007)
18. Kumari, P., Pretschner, A., Peschla, J., Kuhn, J.: Distributed data usage control for web applications: a social network implementation. In: Proc. 1st ACM Conf. on Data and Application Security and Privacy, pp. 85–96 (2011)
19. Pretschner, A., Lovat, E., Büchler, M.: Representation-Independent Data Usage Control. In: Garcia-Alfaro, J., Navarro-Arribas, G., Cuppens-Boulahia, N., de Capitani di Vimercati, S. (eds.) DPM 2011 and SETOP 2011. LNCS, vol. 7122, pp. 122–140. Springer, Heidelberg (2012)
20. Feth, D., Pretschner, A.: Flexible Data-Driven Security for Android. In: SERE 2012, pp. 41–50 (June 2012)
21. Kumari, P., Kelbert, F., Pretschner, A.: Data Protection in Heterogeneous Distributed Systems: A Smart Meter Example. In: INFORMATIK 2011 - Dependable Software for Critical Infrastructures (2011)
22. Kumari, P., Pretschner, A.: Deriving implementation-level policies for usage control enforcement. In: Proc. 2nd ACM Conference on Data and Application Security and Privacy, CODASPY 2012, pp. 83–94. ACM (2012)
23. Spivey, J.M.: The Z Notation: A Reference Manual. Prentice Hall, UK (1998)
24. Roque, R.: Open Blocks (2009), http://education.mit.edu/openblocks
25. Abadi, M., Lamport, L.: The existence of refinement mappings. In: LICS 1988 (1988)
26. Su, L., Chadwick, D., Basden, A., Cunningham, J.: Automated decomposition of access control policies. In: Proc. 6th IEEE Intl. Workshop on Policies for Distributed Systems and Networks, pp. 6–8 (2005)
27. Young, J.: Commitment analysis to operationalize software requirements from privacy policies. Requirements Engineering 16, 33–46 (2011)
28. Bandara, A.K., Lupu, E.C., Moffett, J., Russo, A.: A goal-based approach to policy refinement. In: Proc. 5th IEEE Workshop on Policies for Distributed Systems and Networks, pp. 229–239 (2004)
29. Udupi, Y.B., Sahai, A., Singhal, S.: A classification-based approach to policy refinement. In: Proc. 10th Intl Symp. on Integrated Network Management (2007)
30. Davy, S., Jennings, B., Strassner, J.: Conflict Prevention Via Model-Driven Policy Refinement. In: State, R., van der Meer, S., O'Sullivan, D., Pfeifer, T. (eds.) DSOM 2006. LNCS, vol. 4269, pp. 209–220. Springer, Heidelberg (2006)
31. Basile, C., Lioy, A., Scozzi, S., Vallini, M.: Ontology-Based Policy Translation. In: Herrero, Á., Gastaldo, P., Zunino, R., Corchado, E. (eds.) CISIS 2009. AISC, vol. 63, pp. 117–126. Springer, Heidelberg (2009)
32. Guerrero, A., Villagrá, V.A., de Vergara, J.E.L., Sánchez-Macián, A., Berrocal, J.: Ontology-Based Policy Refinement Using SWRL Rules for Management Information Definitions in OWL. In: State, R., van der Meer, S., O'Sullivan, D., Pfeifer, T. (eds.) DSOM 2006. LNCS, vol. 4269, pp. 227–232. Springer, Heidelberg (2006)
33. Aziz, B., Arenas, A.E., Wilson, M.: Model-Based Refinement of Security Policies in Collaborative Virtual Organisations. In: Erlingsson, Ú., Wieringa, R., Zannone, N. (eds.) ESSoS 2011. LNCS, vol. 6542, pp. 1–14. Springer, Heidelberg (2011)
34. Craven, R., Lobo, J., Lupu, E., Russo, A., Sloman, M.: Decomposition techniques for policy refinement. In: Proc. CNSM 2010, pp. 72–79 (2010)

35. O'Rourke, C., Fishman, N., Selkow, W.: Enterprise architecture using the Zachman Framework. Course Technology (2003)
36. Zachman, J.A.: A framework for information systems architecture. IBM Syst. J. 26, 276–292 (1987)
37. The Open Group. TOGAF Version 9 (2009)
38. Gruler, A., Harhurin, A., Hartmann, J.: Modeling the functionality of multi-functional software systems. In: Proc. ECBS 2007, pp. 349–358. IEEE Computer Society (2007)
39. Ziegenbein, D., Braun, P., Freund, U., Bauer, A., Romberg, J., Schatz, B.: Auto-mode - model-based development of automotive software. In: Proc. DATE 2005, pp. 171–177. IEEE Computer Society (2005)
40. Penzenstadler, B.: Tackling Automotive Challenges with an Integrated RE & Design Artifact Model. In: Meersman, R., Tari, Z., Herrero, P. (eds.) OTM 2008 Workshops. LNCS, vol. 5333, pp. 426–431. Springer, Heidelberg (2008)
41. Miller, J., Mukerji, J.: Mda guide version 1.0.1. Technical Report omg/03-06-01, Object Management Group (OMG) (June 2003)
42. Pretschner, A., Schütz, F., Schaefer, C., Walter, T.: Policy evolution in distributed usage control. Electr. Notes Theor. Comput. Sci. 244, 109–123 (2009)

Compositional Verification
of Application-Level Security Properties

Linda Ariani Gunawan and Peter Herrmann

Department of Telematics
Norwegian University of Science and Technology (NTNU)
Trondheim, Norway
{gunawan,herrmann}@item.ntnu.no

Abstract. Automatic model checking can be employed to verify that
security properties are fulfilled by a system model. However, since security requirements constrain most, if not all, functional modules of a
system, such a proof needs to consider nearly all of the system's control
and data flows. For complex real-life applications, that leads to a large
state space to be explored effectively restricting the applicability of a
model checker. To deal with this problem, we advocate a compositional
approach utilizing the features of our model-based engineering technique
SPACE. Both functional behavior and security-related aspects are specified using UML 2 activities. Further, we supplement each activity with
an interface behavior description which will be extended by a security
contract modeling certain security properties to be fulfilled by the activity. This enables us to verify application-level security properties by using
contracts instead of their respective activities in model checker runs so
that the number of states to be checked is significantly reduced. The approach is exemplified by an Android application example in which one's
location must only be shared with certain recipients.

1 Introduction

An often underestimated reason for vulnerabilities and risks in application-level
security is that development flaws in real-life software systems are overlooked.
For instance, Iyer et al. [1] found out that 18% of all vulnerabilities listed in
the *Bugtraq* database resulted from design errors. To avoid such development
flaws, we extended our model-based approach SPACE for the development of
reactive systems [2] and its tool-set Arctis [3] to support also the creation of
secure software [4]. Engineering with SPACE and Arctis profits from the fact
that models are a clearer and more concise way to express a system than traditional program code. That makes it easier to keep track of the system behavior.
Moreover, due to its formal semantics [5], one can verify by syntactic inspection
and model checking that application-level security goals are kept by the system
model [6]. Finally, SPACE uses automatic code generation guaranteeing that the
implementation is a correct realization of the model [2]. Thus, we can be sure
that also the executed code complies with the proven security properties.

J. Jürjens, B. Livshits, and R. Scandariato (Eds.): ESSoS 2013, LNCS 7781, pp. 75–90, 2013.
© Springer-Verlag Berlin Heidelberg 2013

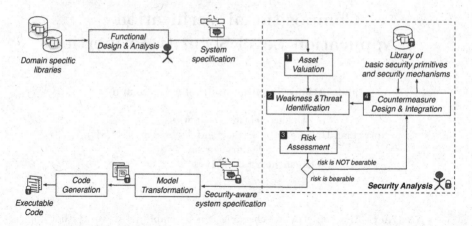

Fig. 1. Security-enhanced Development Method (taken from [4])

While model checking can be executed with a high degree of automatism, its weak point is the state explosion problem [7] which, in effect, constitutes the limiting factor for applying it to large systems. That is especially relevant if one wants to prove security requirements that often define and constrain all functional modules of a system such that its whole state space has to be considered (see [8]).

To tackle the state explosion problem, we advocate compositional verification that is already used to verify properties related to the functionality [3] and reliability [9] of a system. Here, we utilize the model composition mechanism of SPACE in which behavior is specified by an arbitrary number of UML 2 activities [10]. Like Petri-nets, those are graphs modeling behavior as a flow of tokens between the vertices via the edges. Activities are coupled with one another by call behavior actions that we call *building blocks*. From one viewpoint, a building block refers to a particular behavior expressed by an activity. From the other viewpoint, a designator of a block may be incorporated in another activity, and by so-called pins, tokens may flow between activities. Further, a building block is amended with a behavioral interface description specifying the order of token flows through its pins. One advantage of the approach is a high degree of reuse. A block modeling recurrent behavior can be created once and stored in a library. Thereafter, by adding its designator to other activities in a drag and drop fashion, the behavior modeled by the block can easily be added to various system models. According to our experience, on average 70% of a system model corresponds to building blocks taken from assorted libraries [5].

The other advantage of using building blocks is compositional verification [3]. Here, when proving that an activity fulfills a certain property, we can replace the activity of each of its building blocks by the block's behavioral interface description. As the interface description usually models a much simpler functionality than the activity, the number of states to be checked is vastly reduced (see, e.g., [9]). To use compositional verification also for the proof of security properties, however, we have to extend the behavioral interface descriptions by

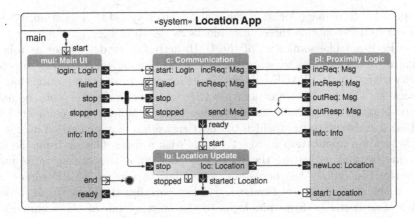

Fig. 2. The Location Application

so-called *security contracts* modeling security properties to be fulfilled by a block. In the model checker runs, we can then use these block-wise security properties to verify that the overall system fulfills the system-wide ones.

Our approach facilitates the cooperation of application domain engineers with security experts (see Fig. 1). In a first step, the domain engineers develop a system specification utilizing blocks from the domain specific libraries. When the model passes all checks for functional correctness [3], it is handed over to the security experts who subject it to a security analysis. The outcome of this analysis is an amended system specification containing only application-level security risks that seem bearable. In a final step, the extended model is automatically transformed to executable code in a two-step process.

In the context of security analysis, the verification of system security properties is used to detect potential flaws in the design which make the system vulnerable against malicious attacks threatening its assets. Of course, such flaws form a formidable risk for the system and the obvious countermeasure is to change the system model such that its behavior fulfills the security properties.

2   Location Application – An Example

The system specification of our example is depicted in Fig. 2. It is an Android application that allows one to share one's current location, but only to a set of intended recipients, i.e., friends. The specification consists of four building blocks implementing various functionalities. The graphical user interface (UI) block *mui: Main UI* handles the user's input and displays relevant information for the user on the device interface. Block *c: Communication* handles the exchange of messages with peer applications running on other devices. This block encapsulates the *XMPP Client Android* block which allows one to transmit messages through an XMPP server. The current location of a device executing this application is reported by block *lu: Location Update*. Finally, block *pl: Proximity Logic* is responsible to manage location sharing, e.g., to respond to a location request from a friend. It contains three inner blocks as shown in Fig. 3.

The Petri-net like semantics of the UML activities models states as tokens resting in token places and state transitions as moving tokens along directed activity edges [10]. In SPACE, all behavior follows the run-to-completion characteristics [5]. This means, transitions are triggered by observable events, namely, the reception of signals and the expiration of local timers, and completed by reaching a stable state from which the next transition may be carried out.

The location application in Fig. 2 begins with a token flowing from the initial node (●) and activating the UI block. Thereafter, the system waits until the device user enters the necessary credentials to use an XMPP server. As soon as the credentials are received, a token carrying the data (in an object of type *Login*) moves from pin *login* to the starting pin of the communication block. Upon successful login, the application proceeds by initiating block *lu: Location Update* to obtain the present geographical location. Subsequently, a token carrying the location data emits from pin *started* of block *lu* and passes through a fork node which duplicates the token. One token is directed to pin *start* of block *pl: Proximity Logic*, while the other one is forwarded to pin *ready* of the UI block. Updates on the current position are reported by block *lu* via pin *loc* and consumed by block *pl* through pin *newLoc*.

The specification shown in Fig. 2 also includes behavior that handles unsuccessful credential verification and an input from the user to stop the application. However, for brevity we do not detail this here, but rather focus on the location sharing functionality which is handled mainly by block *Proximity Logic*.

As depicted on the left side of Fig. 3, block *Proximity Logic* becomes active when a token carrying location data flows from parameter node *start* and passes through a fork node. The downward pointing edge leaving the fork node shows that a token with the location data initializes block *h: Message Handler*. The other outgoing edge indicates that the Java operation *getFriendList* is executed. The output of this operation is a list of friends which is stored locally. This list is forwarded to a fork node with three outgoing edges, one of which initializes block *g: Req Generator*. The second one sends the list of friends to block *h* while the third directs a token through a merge node (◇) to a timer which is started. When the timer expires, block *g* generates location requests, one for each friend in the list, and emits them one-by-one via pin *aReq*. A token flowing through pin *done* indicates that all requests have been yielded and the next batch of requests can be generated when the timer expires again.

The inner block *b: Reactive Buffer* decouples message reception from message handling and, hence, is used to buffer messages while block *h* is busy processing one message. A message is received through pin *add* of the buffer. When the buffer is empty, it is emitted immediately via pin *out*; otherwise, it is buffered. Invoking pin *next* will get either the subsequent message in the buffer (via pin *out*) or an indication that the buffer is empty (pin *empty*). Three types of messages are buffered and handled, namely, generated requests, requests from peer applications running in different devices, and responses to the generated requests.

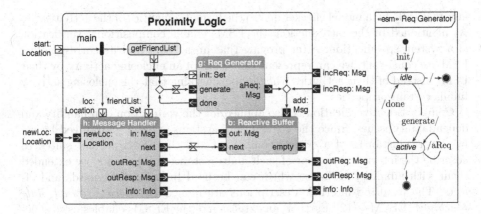

Fig. 3. Block Proximity Logic

A message is received by the message handler block via pin *in*. Depending on the message type and additional constraints, one of the following four alternative behaviors is taken: (1) If the message is a generated request, it is emitted via pin *outReq*. (2) If the message is a request from a person in the friend list, a response containing the latest location is created and emitted via pin *outResp*. (3) If the message is a response to a generated request and the friend's location is near, a notification is emitted via pin *info* (4) For all other cases, the message is dropped. In addition, a token is emitted via the output pin *next* which after a certain latency guaranteed by a timer leads to obtaining the subsequent message from the buffer. The flows via the pins *outReq*, *outResp* and *info* of building block *h* are forwarded to the pins of the same name of the block *Proximity Logic* such that outgoing requests and responds are further sent to the communication block while notifications are forwarded to the UI block (see Fig. 2).

3   Interface Contracts

Except for system-level blocks like the one in Fig. 2, building blocks are supplemented with behavioral interface descriptions. As modeling technique for the interface behavior, we use so-called External State Machines (ESMs) [11] that specify the possible ordering of events visible on the activity pins. The ESM of the block *Req Generator* is depicted on the right side of Fig. 3. It shows that this block starts by receiving a token through pin *init* and entering state *idle*. Thereafter, the block can receive a token via pin *generate* upon which it will emit requests, one at a time, via pin *aReq*. After having generated requests to a list of recipients, the block returns to state *idle* emitting a token via pin *done*. Later, the next batch of requests can be created upon receiving a new *generate* event. The transitions labeled with / show that block *Req Generator* allows its surrounding block, in our case the *Proximity Logic*, to terminate it anytime.

An ESM must be respected both by the activity and its environment in order to guarantee a correct interaction between them. Such property can be verified automatically by a model checker due to the formal semantics of the activities [5]. As mentioned in the introduction, the ESMs enable compositional verification of a system specification: After proving that an activity and its corresponding ESM are consistent, we can represent the blocks of an enclosing activity by their ESMs instead of their activities when model checking that the enclosing activity fulfills certain properties.

Compositional verification is also applied for the verification of reliability and dependability issues. Since the reliability of systems is often guaranteed by using several instances of a critical component and the ESMs are not suited to describe the interface behavior of such multi-instance components, we extended them with auxiliary variables which can be used in transition guards and effects. The resulting interface descriptions are named *Extended External State Machines* (EESMs) [12]. Further, an extension of the EESMs enables us to specify indeterministic interface behavior following from component failures, e.g., non-responsiveness or a reset to the initial state. In consequence, we could reduce the number of states to be model checked by several orders of magnitude (see [9]). This encouraging result has lead us to use compositional verification also for security properties which will be discussed in the following.

4   Modeling Security-Relevant Aspects

A highly relevant asset of applications running on modern smartphones is the phone's location which can be retrieved by the built-in GPS receiver or by triangulation of WiFi base stations. Of course, the location data must not leak to unauthorized principals since that would violate the privacy of the phone user and might also be a severe risk for her/his personal safety. Thus, with respect to application-level security we have to avoid that an erroneous system layout may lead to the unauthorized transmission of the location information. In the example presented in Sect. 2, for instance, we have to guarantee a security property $\mathcal{P}$ expressing that *"one's geographical position may only be sent to one's friends"*.

As described in [3], the semantics of the SPACE approach and its tool-set Arctis is based on Leslie Lamport's *Temporal Logic of Actions* (TLA) [13]. This enables us to specify security properties like $\mathcal{P}$ by abstract system specifications or invariants in TLA and use the model checker TLC [14] to verify that they are fulfilled by the TLA representation of a SPACE model.

A suitable notation for the security contracts used to add security properties to the interface contracts are the EESMs [12]. They allow to insert additional variables and constants in transition guards and effects. As an example, we list the EESM of the block *Proximity Logic* in Fig. 4. It uses three control states, i.e., the initial state (•), *_idle* and *pL_active*. Besides of the pin identifiers similar to those used in the ESMs (see Sect. 3), a transition can be provided by a guard consisting of a logical predicate framed by square brackets as well as operations on the variables which are described using Java-like statements in lined boxes. The EESM in Fig. 4 contains a variable *v_loc* storing the current location of the own device. The initial transition is carried out during system start and leads

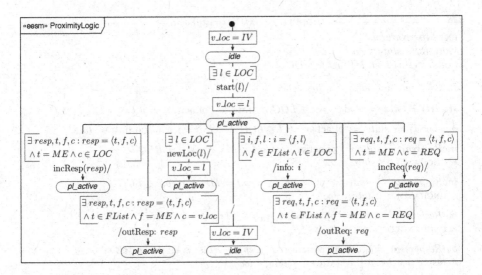

Fig. 4. Security Contract expressed as an EESM

from the initial state to _idle. In its effect part, the variable v_loc is set to an initial value expressed by the constant IV. The activation of the block takes place when a token containing the current location as a parameter l reaches the pin *start*. The corresponding transition switches the control state from *idle* to *pl_active*. Further, it demands that l is indeed location information which is described in the transition guard and sets the variable v_loc to l. The block can only be terminated implicitly by closing down the overall system which, like in the ESMs, is expressed by the transition /. Here, the control state is set back to *idle* again and v_loc to the initial value IV.

Particularly interesting for the security proof of property $\mathcal{P}$ are the transitions *outResp* and *outReq* since the tokens leaving through them contain the messages to be sent by the communication block. The transition *outResp* uses a parameter *resp* specifying the message to be sent in the token. According to the guard of the transition, *resp* is a triple containing the address from the friends list expressed by the constant $FList$ as the recipient address t (*to*). The sender address f (*from*) contains the user's address which is described by the constant ME while the content c includes the device's location data which is stored in the variable v_loc. The transition *outReq* is similar with the exception that the message content is a request (expressed by the constant REQ) asking the recipient for its geographical position. Thus, the EESM specifies that all messages passing pins *outResp* and *outReq* have a friend as a recipient address.

EESMs can be automatically transformed into specifications in TLA$^+$ [13], the notation of TLA and the input language of the model checker TLC (see [12]). TLA is a linear-time temporal logic in which state transition systems are specified using variables for the states and actions (i.e., predicates on pairs of states) for the transitions. The TLA$^+$ specification of the EESM of the block *Proximity Logic* is listed in Fig. 5. It uses the variables *state* denoting the current control

─────────── MODULE *EESMProximityLogic* ───────────

EXTENDS *Naturals*

VARIABLES *state*, *v_loc*

CONSTANTS *FList*, *ME*, *REQ*, *LOC*, *IV*

$Init \stackrel{\Delta}{=} state =$ "_idle" $\wedge$ $v\_loc = IV$

$start(l) \stackrel{\Delta}{=} state =$ "_idle" $\wedge$ $l \in LOC$ $\wedge$ $state' =$ "pl_active" $\wedge$ $v\_loc' = l$

$newLoc(l) \stackrel{\Delta}{=} state =$ "pl_active" $\wedge$ $l \in LOC$ $\wedge$ $v\_loc' = l$ $\wedge$ UNCHANGED *state*

$incReq(req) \stackrel{\Delta}{=} state =$ "pl_active" $\wedge$ $req.t = ME$ $\wedge$ $req.c = REQ$
$\wedge$ UNCHANGED $\langle state, v\_loc \rangle$

$incResp(resp) \stackrel{\Delta}{=} state =$ "pl_active" $\wedge$ $resp.t = ME$ $\wedge$ $resp.c \in LOC$
$\wedge$ UNCHANGED $\langle state, v\_loc \rangle$

$outReq(req) \stackrel{\Delta}{=} state =$ "pl_active" $\wedge$ $req.t \in FList$ $\wedge$ $req.f = ME$ $\wedge$ $req.c = REQ$
$\wedge$ UNCHANGED $\langle state, v\_loc \rangle$

$outResp(resp) \stackrel{\Delta}{=} state =$ "pl_active" $\wedge$ $resp.t \in FList$ $\wedge$ $resp.f = ME$
$\wedge$ $resp.c = v\_loc$ $\wedge$ UNCHANGED $\langle state, v\_loc \rangle$

$implicit\_termination \stackrel{\Delta}{=} state =$ "pl_active" $\wedge$ $state' =$ "_idle" $\wedge$ $v\_loc' = IV$

Fig. 5. Security Contract expressed as a TLA+ specification

state of the EESM as well as the additional variable *v_loc*. *Init* is a predicate specifying the beginning state of the block, i.e., *_idle*. The other seven definitions model the transitions of the EESM in form of actions. Here, a simple variable identifier refers to the state before carrying out an action, whereas an identifier marked by a prime symbol (') points to the state after its execution. For example, before triggering the action *start*, the variable *state* is equal to *_idle* while afterwards it carries the value *pl_active*. Further, this action is only enabled if its parameter *l* is of type *LOC* and the variable *v_loc* in the next state is *l*.

5 Compositional Verification

In TLA, the verification, that an application specification *Spec* fulfills a security property $\mathcal{P}$, corresponds to the implication proof $Spec \Rightarrow \mathcal{P}$. To carry out this proof, we transform the SPACE model of the application into TLA$^+$ specifications of the activities and EESMs that are coupled with each other in a constraint-oriented way (see [15]), forming the system specification *Spec*. The security property $\mathcal{P}$ is modeled as an abstract system specification or an invariant in TLA$^+$ as well. As discussed above, a system-level activity can contain any number of building blocks referring to other activities which in turn may encapsulate other activities (see Fig. 2 and Fig. 3 as an example). Reflecting that constraint-oriented composition corresponds to conjoining TLA formulas, *Spec* is defined as the conjunction of the TLA$^+$ specifications of all activities modeling the application:

$$Spec \triangleq \mathcal{A}_s \wedge \bigwedge\nolimits_{b \in Blocks} \mathcal{A}_b \tag{1}$$

Here, $Blocks$ is the set of all building blocks, while $\mathcal{A}_b$ denotes the TLA$^+$ specification of the activity referenced by block b. With $\mathcal{A}_s$, we refer to the system activity. For our location application in Fig. 2, $\mathcal{A}_s$ corresponds to activity $Location\ App$ while the set $Blocks$ contains eight elements, namely, mui, c, lu, pl, h, b, g, and x. The first four elements refer to the activities $Main\ UI$, $Communication$, $Location\ Update$, and $Proximity\ Logic$ respectively (see Fig. 2). The elements h, b and g point to the activities enclosed by the block pl (see Fig. 3), while x marks the activity $XMPP\ Client$ which is enclosed by the communication block c.

To prove $Spec \Rightarrow \mathcal{P}$ by compositional verification, we have to conduct two major steps. First, we verify that all activities $\mathcal{A}_b$ (except the one on system level) are consistent with their corresponding EESMs. To clarify this proof, we perceive a system specification as a tree of activities. Here, an activity $\mathcal{A}_b$ is the parent of another activity $\mathcal{A}_c$ if the designator of the building block c referring to $\mathcal{A}_c$ is enclosed in $\mathcal{A}_b$. The system activity $\mathcal{A}_s$ forms the root of this tree, while those activities not containing any building blocks are the leaves.

We prove now for every activity $\mathcal{A}_b$ in the tree except for the root that it fulfills its EESM $\mathcal{E}_b$ whereupon we represent its children activities by their EESMs:

$$\mathcal{A}_b \wedge \bigwedge\nolimits_{c \in Children(b)} \mathcal{E}_c \Rightarrow \mathcal{E}_b \tag{2}$$

Proving equation (2) for all activities except for the system activity is sufficient since one can deduce by induction that all activities fulfill also the equation

$$\mathcal{A}_b \wedge \bigwedge\nolimits_{c \in Descendants(b)} \mathcal{A}_c \Rightarrow \mathcal{E}_b \tag{3}$$

in which $Descendants$ refer to all the descendants of an activity in the tree. The starting step of the induction is the verification that equation (3) follows from (2) for all leaves of the tree. This proof is trivial since the leaf activities do not have any descendants at all. In the inductive step, we have to verify that an activity $\mathcal{A}_b$ fulfilling equation (2) also guarantees equation (3) as long as (3) holds also for all of its children. Likewise, this proof is easy since the descendants of the children of $\mathcal{A}_b$ are also its own descendants. Therefore,

$$\forall k \in Children(b) : \mathcal{A}_b \wedge \bigwedge\nolimits_{c \in Descendants(b)} \mathcal{A}_c \Rightarrow \mathcal{E}_k$$

holds and equation (3) can be directly deduced from (2).

In TLA, a verification of equation (2) is achieved by employing a refinement mapping [16], i.e., a mapping between the state spaces of $\mathcal{A}_b$ and $\mathcal{E}_b$ guaranteeing that an initial state of $\mathcal{A}_b$ is mapped to an initial state of $\mathcal{E}_b$, and that a TLA action of $\mathcal{A}_b$ is either mapped to an action in $\mathcal{E}_b$ or to a stuttering step in which the variables in $\mathcal{E}_b$ do not change. The refinement mapping proof can be automated by the model checker TLC, whereat the use of the EESMs of $\mathcal{A}_b$'s children keeps the number of states to check low. We cannot detail the verification process here, but the proofs are similar to the ones presented in [12].

One of the EESM proofs in our location example was $\mathcal{A}_{pl} \wedge \mathcal{E}_g \wedge \mathcal{E}_h \wedge \mathcal{E}_b \Rightarrow \mathcal{E}_{pl}$ stating that the activity *Proximity Logic* in Fig. 3 fulfills its EESM that is depicted in Fig. 4.

In the second major proof-step, we use the EESMs of the children of the root activity $\mathcal{A}_s$ to verify the security property $\mathcal{P}$:

$$\mathcal{A}_s \wedge \bigwedge_{c \in Children(s)} \mathcal{E}_c \Rightarrow \mathcal{P} \qquad (4)$$

From this equation and the fact that the children of $\mathcal{A}_s$ are blocks in the system specification *Spec*, we can infer $Spec \Rightarrow \mathcal{P}$ since for all the children of $\mathcal{A}_s$ equation (3) holds as well. For the proof of equation (4), we use the model checker TLC which again profits from using the EESMs of the inner blocks instead of their activities such that the number of states to be checked can be reduced in all of our TLC model checker runs.

An excerpt of the TLA$^+$ specification of our location application is depicted in Fig. 6, in particular, $\mathcal{A}_{LocationApp} \wedge \mathcal{E}_{mui} \wedge \mathcal{E}_c \wedge \mathcal{E}_{lu} \wedge \mathcal{E}_{pl} \Rightarrow \mathcal{P}$. Variables and constants used in the specification are declared in the section *Variables and Constants Declaration*. Most of them represent the variables and constants defined in the EESMs, including the ones modeling security related aspects.

—— MODULE *Location_App* ——

Variables and Constants Declaration

VARIABLES $pl\_state, c\_state, mui\_state, lu\_state, pl\_v\_loc, c\_v\_out, c\_v\_enOut, c\_v\_enLgn$
CONSTANTS $FList, ME, REQ, LOC, IV, Any, Ciphertext, Login$

Using the EESMs of Inner Blocks

$mui \triangleq$ INSTANCE $EESMMainUI$ WITH $state \leftarrow mui\_state$
$c \triangleq$ INSTANCE $EESMCommunication$ WITH $state \leftarrow c\_state, v\_out \leftarrow c\_v\_out,$
$v\_enOut \leftarrow c\_v\_enOut, v\_enLgn \leftarrow c\_v\_enLgn, Recepient \leftarrow FList \cup Any$
$lu \triangleq$ INSTANCE $EESMLocationUpdate$ WITH $state \leftarrow lu\_state$
$pl \triangleq$ INSTANCE $EESMProximityLogic$ WITH $state \leftarrow pl\_state, v\_loc \leftarrow pl\_v\_loc$

System Actions

$pl\_outResp(resp) \triangleq pl!outResp(resp) \wedge c!send(resp)$
$\wedge$ UNCHANGED $\langle mui\_state, lu\_state \rangle$
$\ldots$

TLA$^+$ System Specification

$Init \triangleq pl!Init \wedge c!Init \wedge mui!Init \wedge lu!Init$
$Next \triangleq$
$\vee \exists r \in [t:FList \cup Any, f:\{ME\} \cup Any, c:\{REQ\} \cup LOC \cup Any] : pl\_outResp(r)$
$\vee \ldots$
$vars \triangleq \langle pl\_state, c\_state, mui\_state, lu\_state, pl\_v\_loc, c\_v\_out, c\_v\_enOut, c\_v\_enLgn \rangle$
$Spec \triangleq Init \wedge \Box[Next]_{\langle vars \rangle}$

$\mathcal{P} \triangleq \Box((c\_v\_out = IV) \vee (c\_v\_out.c \in LOC \Rightarrow c\_v\_out.t \in FList))$

Fig. 6. Excerpt from the TLA+ specification of the Location App

The section *Using the EESMs of Inner Blocks* contains four instantiation statements used to couple TLA$^+$ specifications into the system description. For instance, in the last statement, module *EESMProximityLogic* (see Fig. 5) is instantiated and denoted as *pl*. Here, the variables *state* and *v_loc* of the instantiated module are respectively substituted by variables *pl_state* and *pl_v_loc* of the application specification. Likewise, all the constants in the EESM are instantiated, albeit implicitly since they are substituted with constants of the same name. By the other three statements, the EESMs of the block instances *mui*, *c*, and *lui* are composed. The instances enable us to refer to EESM transitions by $\langle instance \rangle ! \langle EESM\_transition \rangle$ (e.g., $pl!outResp(resp)$). These references are used to specify events in an enclosing activity as exemplified in Fig. 6 by the section labeled with *System Actions*. The TLA action $pl\_outResp(resp)$ defines that response *resp* emitted by block *Proximity Logic* ($pl!outResp(resp)$) is sent by block *Communication* ($c!send(resp)$) which, among others, stores a message sent to another station in the auxiliary variable[1] c_v_out. Moreover, the UNCHANGED statement points out that the blocks *Main UI* and *Location Update* are not involved in the action and do not change their variables.

In section *TLA$^+$ System Specification* of Fig. 6, the TLA$^+$ specification modeling block *Location App* is written as the so-called canonical formula $Spec \triangleq Init \wedge \square[Next]_{\langle vars \rangle}$. It expresses that the initial state of the application fulfills the predicate *Init* and that every state change follows one of the system actions which are disjuncts of the next state relation *Next*. By $[\ldots]_{\langle vars \rangle}$, one models that stuttering steps in which the list of variables *vars* do not change are also allowed.

The security property $\mathcal{P}$, i.e., *"one's locations are only sent to one's friends"* is expressed by the TLA invariant that is listed in the bottom part of Fig. 6. It states that at all times the variable c_v_out storing the messages sent to other recipients carries either the initial value *IV* (i.e., no message has been sent yet) or that a sent message containing location information is sent to the address of a friend. We use the TLC model checker to verify both, $\mathcal{P}$ and the security property *"the credentials used to login to an XMPP server are sent via a secure communication channel"*. The performance issues of the model checker runs proving these two security properties will be discussed below.

6   Model Checking Performance

To evaluate the advantage of employing EESMs for verification of security properties, we compare the result of model checking the example application with two approaches: The first one is the compositional technique described above, i.e., proving formula (4) for the system block *Location App* in Fig. 2 and formula (2) for the eight inner blocks *mui*, *c*, *lu*, *pl*, *h*, *b*, *g*, and *x*. The other one is the direct method in which the TLA$^+$ specifications of all activities of the system are used, i.e., equation (1).

---

[1] Auxiliary variables do not influence the behavior of a system but support verification.

Table 1. Verification effort: compositional approach vs. direct approach

		Number of elements for each set			
		1	2	3	4
compositional approach	states (largest)	984	53 855	823 174	6 568 677
	states (total)	1 204	56 200	837 810	6 630 076
	time (total)	2 sec	12 sec	44 sec	331 sec
direct approach	states	6 248	1 047 503	> 25 M	-
	states (x largest)	6.35 x	19.45 x	> 30 x	-
	states (x total)	5.19 x	18.64 x	> 29 x	-
	time	6 sec	224 sec	> 2 hours	-
	time (x total)	3 x	18.67 x	> 163 x	-

• buffer size = 2

We model checked the compositional and direct approaches on a 2.4 GHz, 8 GB personal computer. The result is presented in Tab. 1. Both versions use the same sets representing various types of data (e.g., *FList* denotes a list of friends). Since TLC works by generating behaviors that satisfy a specification, we needed to declare the elements of those sets. We used the same elements for each set in the specifications of both approaches and decided to use the size of a set as the parameter to compare the verification effort. Further, one restriction, i.e., a maximum buffer size, was required since, otherwise, the specifications related to the reactive buffer would have infinitely many reachable states due to arbitrarily many sequences of messages stored by the buffer. For the compositional approach, we present three types of data in Tab. 1: The values in the first row, obtained from proving formula (2) for block *pl: Proximity Logic*, show the largest number of states created in a single model checker run reflecting the maximum amount of memory needed. In the second row, we simply add the number of states checked in all nine runs. Similarly, the total amount of time to model check all nine blocks is shown in the third row. These values are compared with the respective number of states and verification time of the direct approach.

Observing Tab. 1, we see that the number of states found by the model checker for the direct version is much higher compared to the compositional version. In consequence, also the execution time of the model checker runs grew. For example, using sets consisting of two elements, the state space of the direct version is more than 18 times larger than the composed version's. Further, it also takes about 18 times longer to verify the direct version than the compositional one. Moreover, the state and time differences between the compositional and the direct verification increase with a growing set size. Indeed, the direct approach effectively fails when the set size reaches the value 4 while the compositional verification is still manageable in a few minutes.

Altogether, these results confirm our experience with functional and reliability checks mentioned above that utilizing the SPACE building blocks and their interface descriptions for model checking effectively reduces the state explosion problem. Thus, it helps to make automatic analysis more feasible for real-life systems. In addition, the effort to verify systems that are developed with already proven blocks is further reduced since ensuring the conformity of a block to its interface contract only needs to be done once.

7   Related Work

Various methods have been proposed to support the development of secure systems. UMLsec [17] is a UML profile that is used to incorporate security-related information such as fair exchange and secure communication links in various UML diagrams. SecureUML [18] is a modeling language tailored to integrate Role-Based Access Control policies into application models defined with the UML. Similarly, integration of Mandatory Access Control with UML is proposed in [19]. Approaches based on aspect-orientation, modeling security mechanisms as aspects which are automatically weaved in at joint points of a specification, have also been proposed (see, e.g., [20,21,22,23]). The CORAS approach [24] defines a modeling language to support security risk analysis for systems designed with UML. Its UML diagrams are mainly devoted to model the various steps of a security analysis while the purpose of ours is to express system behavior.

Since systems are usually composed from numerous parts, specifying security aspects in the components and verifying system-wide security properties has been the focus of a number of approaches. To support the development of security-critical applications, Moebius et al. proposed SecureMDD, a model-driven technique that includes verification of application-specific properties [25]. For large systems, they take an incremental approach for which some functionality is added in every step such that security proof needs to be repeated in every iteration [26]. In contrast, in our approach functional behaviors are composed in a constraint-oriented way. Deng et al. proposed a method to model security system architectures and verify whether required security constraints are assured by the composition of the components [27]. However, unlike our work, it employs a top-down approach. Security policies are specified as application-wide constraint patterns which are further decomposed onto the individual components of the system. In [28], Khan et al. present a framework to construct compositional security contracts based on the required and ensured security properties exposed by the atomic components. Although the contracts help engineers to characterize the security aspects of a composed system, the framework does not include validation whether the contracts fulfill the security requirements of the system. Other work that aims to address security issues in software systems consisting of simpler components can be found in [29,30,31].

8   Conclusion

In this paper, we introduced *security contracts* that encapsulate both functional and security aspects of a building block. Due to the formal semantics of the contracts, model checking can be employed to ensure that the contracts and their corresponding blocks are consistent. Furthermore, we showed that such contracts enable compositional verification of application-level security properties which significantly reduces the number of states to be checked and, consequently, also the verification time. On the whole, these behavior and security interface descriptions facilitate model-based development of secure systems: Security mechanisms enclosed in building blocks [4,6] are easily integrated with blocks modeling other functionalities, and both kinds of blocks can be (re-)used in various application designs. The security contracts specify security properties fulfilled by the blocks.

Currently, we are investigating in separating the development of the security contracts from using them in proofs of system-wide security properties. This supports the nature of SPACE-based system engineering that building blocks are often developed independently from the applications and stored in libraries. To achieve that, we need to find a way that relevant security aspects of a block can be·anticipated, modeled in a security contract, and proven without knowing the applications using the block. As a solution, we consider to employ application domain-oriented information security ontologies stating relevant assets, vulnerabilities and threats (see also [32]). For example, an ontology for Android may contain passwords and location information as typical assets of Android devices and leaking them as a typical confidentiality threat. Based on that, a security expert might annotate the building blocks of the Android library by security contracts addressing the elements of this ontology. Moreover, an ontology may contain a list of system-wide security properties to be fulfilled by systems of that domain (e.g., an Android device may never send a password to anybody).

Like the security contracts which are represented by EESMs, one can also define the system-wide security properties in a more comprehensible syntax than plain TLA$^+$ effectively reducing the required expertise in formal methods. This complements our experience with functional system development in which engineers analyze their models by just pushing a button which leads to a message containing that everything is correct or a list of errors in an easily understandable format. Further, the trace towards a state violating a property is animated directly on the SPACE models [3] such that the engineer does not need to understand the formalism of the model checker running in the background at all. We want to achieve a similar procedure for the verification of security properties. By describing the system security properties in an easily understandable way, at least basic security protection can be done directly by the domain engineers without involving the security experts in excess of the creation of the building block security contracts. This will ease the development of more secure software.

References

1. Iyer, R.K., Chen, S., Xu, J., Kalbarczyk, Z.: Security Vulnerabilities - from Data Analysis to Protection Mechanisms. In: Proceedings of the Ninth IEEE International Workshop on Object-Oriented Real-Time Dependable Systems, WORDS 2003, pp. 331–338 (2003)
2. Kraemer, F.A.: Engineering Reactive Systems: A Compositional and Model-Driven Method Based on Collaborative Building Blocks. PhD thesis, Norwegian University of Science and Technology (August 2008)

3. Kraemer, F.A., Slåtten, V., Herrmann, P.: Tool Support for the Rapid Composition, Analysis and Implementation of Reactive Services. Journal of Systems and Software 82(12), 2068–2080 (2009)
4. Gunawan, L.A., Herrmann, P., Kraemer, F.A.: Towards the Integration of Security Aspects into System Development Using Collaboration-Oriented Models. In: Ślęzak, D., Kim, T.-H., Fang, W.-C., Arnett, K.P. (eds.) SecTech 2009. CCIS, vol. 58, pp. 72–85. Springer, Heidelberg (2009)
5. Kraemer, F.A., Herrmann, P.: Reactive Semantics for Distributed UML Activities. In: Hatcliff, J., Zucca, E. (eds.) FMOODS/FORTE 2010, Part II. LNCS, vol. 6117, pp. 17–31. Springer, Heidelberg (2010)
6. Gunawan, L.A., Kraemer, F.A., Herrmann, P.: A Tool-Supported Method for the Design and Implementation of Secure Distributed Applications. In: Erlingsson, Ú., Wieringa, R., Zannone, N. (eds.) ESSoS 2011. LNCS, vol. 6542, pp. 142–155. Springer, Heidelberg (2011)
7. McMillan, K.L.: Symbolic Model Checking: an Approach to the State Explosion Problem. PhD thesis, Carnegie Mellon University, Pittsburgh, PA, USA (1992)
8. Davis, A.M.: Software Requirements: Objects, Functions and States, 2nd edn. Prentice-Hall, Inc., Upper Saddle River (1993)
9. Slåtten, V., Kraemer, F.A., Herrmann, P.: Towards Automatic Generation of Formal Specifications to Validate and Verify Reliable Distributed Systems: A Method Exemplified by an Industrial Case Study. In: Proceedings of the 10th ACM International Conference on Generative Programming and Component Engineering, pp. 147–156. ACM, New York (2011)
10. Object Management Group: Unified Modeling Language: Superstructure, version 2.3 (May 2010) (formal/2010-05-05)
11. Kraemer, F.A., Herrmann, P.: Automated Encapsulation of UML Activities for Incremental Development and Verification. In: Schürr, A., Selic, B. (eds.) MODELS 2009. LNCS, vol. 5795, pp. 571–585. Springer, Heidelberg (2009)
12. Slåtten, V., Herrmann, P.: Contracts for Multi-instance UML Activities. In: Bruni, R., Dingel, J. (eds.) FMOODS/FORTE 2011. LNCS, vol. 6722, pp. 304–318. Springer, Heidelberg (2011)
13. Lamport, L.: Specifying Systems: The TLA+ Language and Tools for Hardware and Software Engineers. Addison-Wesley Professional (2002)
14. Yu, Y., Manolios, P., Lamport, L.: Model Checking TLA+ Specifications. In: Pierre, L., Kropf, T. (eds.) CHARME 1999. LNCS, vol. 1703, pp. 54–66. Springer, Heidelberg (1999)
15. Herrmann, P., Krumm, H.: A Framework for Modeling Transfer Protocols. Computer Networks 34(2), 317–337 (2000)
16. Abadi, M., Lamport, L.: The Existence of Refinement Mappings. Theoretical Computer Science 82(2), 253–284 (1991)
17. Jürjens, J.: Secure System Development with UML. Springer (2005)
18. Basin, D., Doser, J., Lodderstedt, T.: Model Driven Security: From UML Models to Access Control Infrastructures. ACM Transactions on Software Engineering and Methodology 15(1), 39–91 (2006)
19. Doan, T., Demurjian, S., Ting, T.C., Ketterl, A.: MAC and UML for Secure Software Design. In: Proceedings of the 2004 ACM Workshop on Formal Methods in Security Engineering, FMSE 2004, pp. 75–85. ACM, New York (2004)
20. Georg, G., Ray, I., Anastasakis, K., Bordbar, B., Toahchoodee, M., Houmb, S.H.: An Aspect-Oriented Methodology for Designing Secure Applications. Information and Software Technology 51(5), 846–864 (2009); Special Issue: Model-Driven Development for Secure Information Systems

21. Mouheb, D., Talhi, C., Nouh, M., Lima, V., Debbabi, M., Wang, L., Pourzandi, M.: Aspect-Oriented Modeling for Representing and Integrating Security Concerns in UML. In: Lee, R., Ormandjieva, O., Abran, A., Constantinides, C. (eds.) SERA 2010. SCI, vol. 296, pp. 197–213. Springer, Heidelberg (2010)

22. Jürjens, J., Houmb, S.H.: Dynamic Secure Aspect Modeling with UML: From Models to Code. In: Briand, L.C., Williams, C. (eds.) MoDELS 2005. LNCS, vol. 3713, pp. 142–155. Springer, Heidelberg (2005)

23. Jézéquel, J.M.: Model Driven Design and Aspect Weaving. Software and System Modeling 7(2), 209–218 (2008)

24. Lund, M.S., Solhaug, B., Stølen, K.: Model-Driven Risk Analysis - The CORAS Approach. Springer (2011)

25. Moebius, N., Stenzel, K., Reif, W.: Formal Verification of Application-Specific Security Properties in a Model-Driven Approach. In: Massacci, F., Wallach, D., Zannone, N. (eds.) ESSoS 2010. LNCS, vol. 5965, pp. 166–181. Springer, Heidelberg (2010)

26. Moebius, N., Stenzel, K., Borek, M., Reif, W.: Incremental Development of Large, Secure Smart Card Applications. In: Proceedings of the 1st Model-Driven Security Workshop, MDSec 2012 (to appear, 2012)

27. Yi, D., Wang, J., Tsai, J.J., Beznosov, K.: An Approach for Modeling and Analysis of Security System Architectures. IEEE Transactions on Knowledge and Data Engineering 15(5), 1099–1119 (2003)

28. Khan, K., Han, J., Zheng, Y.: A Framework for an Active Interface to Characterise Compositional Security Contracts of Software Components. In: Proceedings of the 2001 Australian Software Engineering Conference, pp. 117–126 (2001)

29. Herrmann, P.: Information Flow Analysis of Component-Structured Applications. In: Proceedings of the 17th Annual Computer Security Applications Conference (ACSAC), pp. 45–54. ACM SIGSAC, IEEE Computer Society Press, New Orleans (2001)

30. Mantel, H.: On the Composition of Secure Systems. In: Proceedings of the IEEE Symposium on Security and Privacy, pp. 88–101. IEEE Computer Society (May 2002)

31. Bartoletti, M., Degano, P., Ferrari, G.L.: Security Issues in Service Composition. In: Gorrieri, R., Wehrheim, H. (eds.) FMOODS 2006. LNCS, vol. 4037, pp. 1–16. Springer, Heidelberg (2006)

32. Vasilevskaya, M., Gunawan, L.A., Nadjm-Tehrani, S., Herrmann, P.: Security Asset Elicitation for Collaborative Models. In: Proceedings of the 1st Model-Driven Security Workshop, MDSec 2012 (to appear, 2012)

Towards Verifying Voter Privacy
through Unlinkability

Denis Butin[1], David Gray[2], and Giampaolo Bella[3,4]

[1] Inria, Université de Lyon
INSA-Lyon, CITI-Inria, F-69621, Villeurbanne, France
denis.butin@inria.fr
[2] School of Computing, Dublin City University
Dublin, Ireland
david.gray@computing.dcu.ie
[3] Dipartimento di Matematica e Informatica, Università di Catania, Italy
[4] Software Technology Research Laboratory, De Montfort University, UK
giamp@dmi.unict.it

Abstract. The increasing official use of security protocols for electronic voting deepens the need for their trustworthiness, hence for their formal verification. The impossibility of linking a voter to her vote, often called voter privacy or ballot secrecy, is the core property of many such protocols. Most existing work relies on equivalence statements in cryptographic extensions of process calculi. This paper provides the first theorem-proving based verification of voter privacy and overcomes some of the limitations inherent to process calculi-based analysis. Unlinkability between two pieces of information is specified as an extension to the Inductive Method for security protocol verification in Isabelle/HOL. New message operators for association extraction and synthesis are defined. Proving voter privacy demanded substantial effort and provided novel insights into both electronic voting protocols themselves and the analysed security goals. The central proof elements are described and shown to be reusable for different protocols with minimal interaction.

Keywords: E-voting, Trustworthy Voting System, Privacy, Security Protocols, Formal Methods.

1   Introduction

The use of electronic voting (e-voting) for official elections is on the rise across the world. Security protocols claiming properties that protect voters and guarantee regular elections require formal scrutiny because of their sensitive nature. Voters are asked to trust, in particular, election officials regarding the handling of their votes. With e-voting, they are asked to trust a security protocol with special goals. One key goal of e-voting protocols is to hide the way a particular voter votes. Most recent efforts [17] to advance formal verification are based on process equivalence. Despite substantial progress, issues remain regarding simplification of protocol models or termination of supporting tools.

J. Jürjens, B. Livshits, and R. Scandariato (Eds.): ESSoS 2013, LNCS 7781, pp. 91–106, 2013.
© Springer-Verlag Berlin Heidelberg 2013

The benefits of specifying privacy in an interactive theorem prover have never been explored until now. Isabelle [15], a generic theorem prover, is flexible enough when used with higher-order logic to allow new classes of security properties to be analysed in the framework provided by the Inductive Method [3]. Its extensions for dealing with voter privacy are described and demonstrated on a classical protocol in the sequel of this manuscript. They required new proof techniques and lines of reasoning, whose development in turn demanded substantial effort. Nevertheless, their application to other protocols is expected to be straightforward, as has been the case for the confidentiality argument [14] for example, with most of the proof scripts adapted for new protocols without significant effort. Automated tools are ideal for checking conjectures about protocols quickly. However, the interactive nature of the Inductive Method pays back, also with e-voting, with a greater support to the analyst's understanding of the protocol entanglements than what automated tools offer today.

The most notable findings in this area stem from formalising the protocols with a process algebra and encoding the privacy properties by process equivalence [10]. As detailed below, process equivalence supports a notion of *indistinguishability* between two situations where a voter voted, respectively, for two different candidates. This implies that an observer cannot discern the two situations being formalised. In line with the operational semantics of the protocols specified by the Inductive Method, we develop an operational encoding of privacy based on *unlinkability* of voter with vote, focusing on the associations that an active attacker can derive from intercepting the protocol traffic. For example, if Alice sent her vote for Bob to the election administrator as a clear-text, then the attacker would build the association Alice-Bob.

However, actual protocol messages are complicated nestings of advanced cryptographic operations (which are still assumed to be reliable), so that the attacker's inspection is far from straightforward. This inspection is formalised by the innovative *association analyser* operator *aanalz* — naming is coherent with the existing lingo. Also, the attacker can intelligently merge associations when they have at least an element in common, similarly to an investigator relating Alice to a crime scene because she wears the same shoe size as that of a shoeprint in the scene. This merge is formalised by the innovative *association synthesiser* operator *asynth*. When it is impossible to build, by means of analysis and then synthesis, an association that features both voter and vote, then there is unlinkability of voter with vote, hence the protocol enforces voter privacy (about their vote). Conversely, the protocol violates voter privacy, irrespectively of how many other voters cast that vote.

An outline of the indistinguishability and unlinkability approaches to modelling privacy (§2) leads to our extensions to the Inductive Method to account for privacy specification and analysis (§3). These extensions are then demonstrated on a classical e-voting protocol known as FOO [12] through its inductive specification (§4) and verification of voter privacy (§5). Conclusions and future work end the manuscript (§6).

2   Modelling Privacy

Voter privacy (also known as ballot secrecy) is generally [9,10] defined as follows: how a particular voter voted is not revealed to anyone. Votes may or may not be published at the end of an election, so it is not the confidentiality of the vote in itself that matters but its association with the voter who cast it. In other words, the way a voter votes should not be discoverable by anyone, even after vote count. A caveat on this definition is the exclusion of the corner case where all voters vote identically.

2.1   Indistinguishability

A common way of modelling privacy involves showing the indistinguishability between two situations:

1. Voter V_a votes x and Voter V_b votes y
2. Voter V_a votes y and Voter V_b votes x

Indistinguishability here means that when V_a and V_b swap their votes, no party (including trusted parties running the election) can tell situations 1 and 2 apart.

Formal analysis is often performed in cryptographic extensions of process calculi, with the applied pi calculus [2] being most typical. Automated tools such as ProVerif [7] or, more recently, AKiSs [9] can be used to assist with such analysis.

ProVerif is used to check protocols represented by processes modelled in the applied pi calculus. It does not restrict the number of protocol sessions. A stronger condition than observational equivalence between processes is checked. Since the validity criterion is an under-approximation, spurious attacks may be found in some cases. There is no risk for flawed protocols to be deemed correct, but correct protocols may be invalidated by the tool because of the approximation. Various approaches to checking voter privacy have been presented. Notably, Kremer and Ryan [13] presented an analysis with some manual parts. In 2008 [11], a fully automatic verification was done. However, a translation algorithm was used without formal proof of correctness. The next year, Delaune, Kremer and Ryan published a detailed analysis in which the number of voters is fixed, with a partially automated privacy proof [10]. New cryptographic primitives can be added easily to the tool via equational theories, but the resulting processes may not terminate in some cases.

AKiSs, the most recent automated tool able to check privacy automatically, is also based on equivalence properties. However, a new kind of cryptographic process calculus is used and a different type of process equivalence is checked, called trace equivalence. Under- and over-approximations of trace equivalence are used to detect flawed protocols and validate correct ones, respectively. The set of supported cryptographic primitives is broader than in ProVerif. For a specific class of processes, called *determinate*, a precise verification can be done. However, not all e-voting protocols fall in this class, in which case one of the approximations must be used. The number of sessions must be bounded as it has critical impact on the computational cost.

2.2   Unlinkability

In contrast to the indistinguishability modelling of privacy, an operational view reflects the natural threat model of an attacker monitoring all network traffic and using the data she can extract to associate a voter with a vote. An outline of this approach and comparison with the one based on indistinguishability first appeared in our recent position paper [8]. Initially, the attacker decomposes each individual message, and records all plaintexts and ciphertexts for which keys are available. She can also associate these with the intended recipient agent of the message. For each protocol event whereby an agent sends a message to another agent, this analysis gives the attacker a set of (components of) messages, namely an association. Moreover, if the communication channel is not anonymous, then the attacker can also extend the association just gathered by storing the identity of the sender.

However, it is not sufficient to inspect in isolation each of the messages sent in the traffic. A voter's identity V may appear near an element m that is later to be extracted again, this time in conjunction with the vote N_v. In this case, such a common element m provides the link between voter V and vote N_v. An attacker monitoring the network sees messages as discrete entities and can exploit the shared context of elements extracted from one given message. This process of combining sets of associations builds up an association synthesis. When all possible protocol scenarios are taken into account, establishing voter privacy boils down to inspecting the synthesised set for the presence of a voter's vote.

The only pieces of information that should not be treated as a possible link to synthesise new associations are those that can be linked to *all* voters, such as the name of the precise election officials that a protocol prescribes. Because their identities appear in each and every protocol session, using one of them as a link would lead to the synthesis of insignificant, that is, privacy-irrelevant, associations. For example, an investigator will not call up every human being as suspect of murder simply upon the basis that everyone could pull a trigger. We shall see that with the FOO protocol, the administrator and the collector are omnipresent, hence must be ruled out to synthesise significant associations.

Without setting bounds on the number of agents, sessions, or message nesting depth, the number of different associations that the attacker can synthesise is very large. Precisely, an unbounded number of associations can be derived by observing a full trace, due to the fact that its length is unbounded. This size limits the tool support that traditional finite-state search can offer. As experienced before with other goals [3], inductive reasoning bypasses the size constraints also with the analysis of associations.

3   Specifying Unlinkability in Isabelle/HOL

3.1   Isabelle/HOL and the Inductive Method

Isabelle is a generic interactive theorem prover supporting many logics. The most commonly used one, HOL, allows formalisation and proof of predicates in higher-order logic. Automated reasoning tools are available, but the user must still define the line of reasoning and guide the proving process. A file-based hierarchy of theories is available. All theorems from parent theories are available when those theories are imported in the current one.

The Inductive Method for security protocol verification was introduced in Paulson's paper [16] and later applied widely, notably to electronic payment [5], non-repudiation [6], certified e-mail [3] and multicast protocols [14]. Its central idea is the use of mathematical induction to model security protocols and their properties. The proofs are also done by induction. A specification of the standard Dolev-Yao threat model, common cryptographic primitives and their properties are provided. All seminal elements of the Inductive Method reside in three theory files. *Message* describes messages and agents, *Event* specifies network event datatypes and *Public* contains the lemmas relevant to cryptographic keys and initial states.

Because of the nature of induction, both the number of protocol sessions and participating agents are unbounded. This allows detection of interleaving or replay attacks. The threat model is incarnated by a special agent called *Spy*, who sees all protocol messages, decrypts whatever she can, and participates actively by sending anything she can build from parts previously obtained. The Spy's capabilities are subsumed by an inductive rule called *Fake*:

| *Fake*: [[*evsf* ∈ *ns_public*; X ∈ *synth* (*analz* (*knows Spy evsf*))]]
  ⟹ *Says Spy B X # evsf* ∈ *ns_public*

Protocol steps are also modelled as inductive rules with pre- and postconditions. Security properties are proven by checking that inductive theorem statements hold over all possible network histories (*traces*).

Available network events are *Says*, *Gets* and *Notes*. The latter represents internal storage of a message by an agent. *Says A B X* represents the sending of message X by agent A to agent B. Delivery does not have to happen, but when it does, this is denoted using *Gets*.

The message operators are as follows:

- *analz* formalises the breaking-up of messages without cryptanalysis. Plaintext is only extracted when the relevant decryption key is part of the knowledge of the agent applying the operator.
- *parts* returns all message building blocks ; it can be seen as *analz* expanded with cryptanalysis.
- *synth* applied to a set of message returns the set of compound messages.

Asymmetric cryptography is available through functions *priEK* and *pubEK* for private and public encryption keys, and then *priSK* and *pubSK* for private and public signing keys. Each of them takes the proprietor agent as a parameter. A private key of a given operation mode is required to decrypt a message encrypted with the corresponding public key, and conversely.

Agent knowledge, formalised by the function *knows*, maps an agent and a list of network events to a set of messages: the knowledge that the agent can extract from this trace. Agents already know some elements (those in *initState*) such as keys before a protocol even begins.

A hands-on, step by step guide to using the Inductive Method can be found in a recent paper [4].

3.2 Extensions for Unlinkability

The analysis of associations requires a new message operator, *analzplus*. It is built on the traditional *analz* message operator, endowed with an external message set providing extra decryption keys:

inductive_set
  analzplus :: *msg set* $\Rightarrow$ *msg set* $\Rightarrow$ *msg set*
  for H :: *msg set* **and** *ks* :: *msg set*
  where
  Inj [*intro,simp*]: $X \in H \implies X \in analzplus\ H\ ks$
  | *Fst*: $\{X,Y\} \in analzplus\ H\ ks \implies X \in analzplus\ H\ ks$
  | *Snd*: $\{X,Y\} \in analzplus\ H\ ks \implies Y \in analzplus\ H\ ks$
  | *Decrypt* [*dest*]: $[Crypt\ K\ X \in analzplus\ H\ ks;\ Key\ (invKey\ K) \in analzplus\ H\ ks]$
                $\implies X \in analzplus\ H\ ks$
  | *Decrypt2* [*dest*]: $[Crypt\ K\ X \in analzplus\ H\ ks;\ Key\ (invKey\ K) \in ks]$
                $\implies X \in analzplus\ H\ ks$

In particular, the new operator is useful to formalise everything, namely the set of all message components, that the attacker can extract from a single message sent in the traffic by hammering it with the entire knowledge she has acquired on an entire trace. For a message X and a trace *evs*, this set can be defined as *analzplus* $\{X\}$ (*analz*(*knows Spy evs*)).

Using *analzplus*, the message association analyser *aanalz* can be defined inductively. Only *Says* events influence it. Indeed, each *Gets* message reception event follow a message sending event *Says*, and *Notes* events correspond to private recording of data by agents:

primrec *aanalz* :: *agent* => *event list* => *msg set set*
where
  aanalz_Nil: *aanalz A* [] = {}
| *aanalz_Cons*:
  aanalz A (*ev* # *evs*) =
   (*if A = Spy then*
   (*case ev of*
     Says A' B X $\Rightarrow$
     (*if A'* $\in$ *bad then aanalz Spy evs*
      else if isAnms X
         then insert ({*Agent B*} $\cup$ (*analzplus* {X} (*analz*(*knows Spy evs*))))
                 (*aanalz Spy evs*)
         else insert ({*Agent B*} $\cup$ {*Agent A'*} $\cup$
                 (*analzplus* {X} (*analz*(*knows Spy evs*)))) (*aanalz Spy evs*))
   | *Gets A' X* $\Rightarrow$ *aanalz Spy evs*
   | *Notes A' X* $\Rightarrow$ *aanalz Spy evs*)
   else aanalz A evs)

The definition indicates, among other aspects, that only the attacker can analyse associations. Also, she will neglect the associations created by compromised agents, thus including those that she may have created, by sending out specific messages. It can also be seen that the sender identity is extracted only for messages that are not sent anonymously. The *isAnms* predicate holds of messages with a specific form that we conventionally interpret to signify anonymity.

The association synthesiser *asynth* can be introduced now. Its definition is not tied to *aanalz*, but it will always be used in conjunction with it for our purposes. Specifically, we will examine the contents of the set *asynth (aanalz Spy evs)*, where *evs* is a generic protocol history. The *asynth* operator introduces a new association as the union of association sets that share a common element:

inductive_set
 asynth :: *msg set set* $\Rightarrow$ *msg set set*
 for *as* :: *msg set set*
 where
 asynth_Build [*intro*]: $[\![a1 \in as;\ a2 \in as;\ m \in a1;\ m \in a2;$
 $\qquad\qquad\qquad m \neq Agent\ Adm;\ m \neq Agent\ Col]\!]$
 $\qquad\qquad \Longrightarrow a1 \cup a2 \in asynth\ as$

As noted above, the definition insists that the common element is not a piece of information that can be linked to *all* voters — for instance, the name of election officials since they appear in every step. The version below can be used for protocols that define two election officials, here called *Adm* and *Col*, in line with the subsequent case study.

4   Modelling the FOO Protocol in the Inductive Method

The well-known Fujioka, Okamoto and Ohta (FOO) [12] protocol features two election officials called administrator and collector and involves bit commitments as well as blind signatures. The specification of its protocol steps follows after a description of some extensions.

Blind signatures are a cryptographic primitive often found in e-voting protocols, and in particular in the FOO protocol. We specify them for the first time in the Inductive Method, as an inductive rule in the protocol model. The Spy gains knowledge of a plain signature if she knows the corresponding blinded signature and blinding factor, modelled as a symmetric key:

| *Unblinding*:
 $[\![evsb \in foo;\ Crypt\ (priSK\ V)\ BSBody \in analz\ (spies\ evsb);$
 $BSBody = Crypt\ b\ (Crypt\ c\ (Nonce\ N));\ b \in symKeys;\ Key\ b \in analz\ (spies\ evsb)]\!]$
 $\Longrightarrow Notes\ Spy\ (Crypt\ (priSK\ V)\ (Crypt\ c\ (Nonce\ N)))\ \#\ evsb \in foo$

Anonymous channels are specified by defining a function to replace *Says* when needed. We are conventionally defining an anonymous message by means of a precise message format — the actual message is prepended with a constant number:

consts *anms* :: *nat*
definition *Anms* :: [*agent, agent, msg*] ⇒ *event* **where**
  Anms A B X ≡ *Says A B* ⦃*Number anms, X*⦄

Administrator and collector are introduced as translations *Adm* and *Col* of specific agents. We now turn to the actual protocol steps and their model.

4.1   FOO Protocol Steps and Inductive Protocol Model

The FOO protocol features six phases that give rise to as many protocol steps and corresponding inductive rules.

1. *Preparation*: The voter V picks a vote N_v, builds N_{vc} using the commitment key c, and blinds this vote commitment using the blinding factor b. V then signs the blinded commitment and sends it to the administrator along with V's identity.

   | *EV1*:
     ⟦*evs1* ∈ *foo*; V ≠ *Adm*; V ≠ *Col*; c ∈ *symKeys*; *Key c* ∉ *used evs1*;
      b ∈ *symKeys*; *Key b* ∉ *used evs1*; b ≠c; *Nonce Nv* ∉ *used evs1*⟧
        ⟹ *Says V Adm* ⦃*Agent V, Crypt* (*priSK V*) (*Crypt b* (*Crypt c* (*Nonce Nv*)))⦄
     # *Notes V* (*Key c*) # *Notes V* (*Key b*) # *evs1* ∈ *foo*

2. *Administration*: Upon reception of a signed, blinded commitment, the administrator opens it and checks that the quoted agent name is equal to the signer of the blind signature. If such is the case and the agent has not voted before, the administrator returns the message to V, now signed by the former. The administrator also records V's name.

   | *EV2*:
     ⟦*evs2* ∈ *foo*; V ≠ *Adm*; V ≠ *Col*; *Notes Adm* (*Agent V*) ∉ *set evs2*;
      Gets Adm ⦃*Agent V, Crypt* (*priSK V*) *BSBody*⦄ ∈ *set evs2*;
      BSBody = *Crypt P R*; ∀ *X Y. MPair X Y* ∉ *parts*{*BSBody*}⟧
        ⟹ *Says Adm V* (*Crypt* (*priSK Adm*) *BSBody*)
        # *Notes Adm* (*Agent V*) # *evs2* ∈ *foo*

3. *Voting*: If V obtained the administrator's reply, V unblinds it and sends the resulting plain signature to the collector over an anonymous channel.

   | *EV3*:
     ⟦*evs3* ∈ *foo*; *Says V Adm* ⦃*Agent V*,
       Crypt (*priSK V*) (*Crypt b* (*Crypt c* (*Nonce Nv*)))⦄ ∈ *set evs3*;
       Gets V (*Crypt* (*priSK Adm*) (*Crypt b* (*Crypt c* (*Nonce Nv*)))) ∈ *set evs3*⟧
        ⟹ *Anms V Col* (*Crypt* (*priSK Adm*) (*Crypt c* (*Nonce Nv*))) # *evs3* ∈ *foo*

4. *Collecting*: The collector checks the signature and publishes the enclosed vote commitment N_{vc} on a bulletin board, provided that it was not published before and that all votes have been received.

| $EV4$:
⟦$evs4 \in foo$; $V \neq Adm$; $V \neq Col$; $Says\ Col\ Col\ CX \notin set\ evs4$;
  $Gets\ Col\ \{\!|Number\ anms,\ Crypt\ (priSK\ Adm)\ CX|\!\} \in set\ evs4$;
  $CX = Crypt\ P\ R$; $\forall\ X\ Y.\ MPair\ X\ Y \notin parts\{CX\}$⟧
  $\implies Says\ Col\ Col\ CX\ \#\ evs4 \in foo$

5. *Opening*: Once N_{vc} has appeared on the bulletin board, V sends c over an anonymous channel so that N_v can be revealed.

| $EV5$:
⟦$evs5 \in foo$; $Says\ V\ Adm\ \{\!|Agent\ V$,
  $Crypt\ (priSK\ V)\ (Crypt\ b\ (Crypt\ c\ (Nonce\ Nv)))|\!\} \in set\ evs5$;
  $Gets\ Col\ (Crypt\ c\ (Nonce\ Nv)) \in set\ evs5$; $Key\ c \in analz\ (knows\ V\ evs5)$;
  $c \notin range\ shrK$; $c \in symKeys$⟧
  $\implies Anms\ V\ Col\ (Key\ c)\ \#\ evs5 \in foo$

6. *Counting*: Upon reception of V's key, the collector publishes N_v on the condition that the key be identical to c.

| $EV6$:
⟦$evs6 \in foo$; $Gets\ Col\ \{\!|Number\ anms,\ Key\ c|\!\} \in set\ evs6$;
  $Gets\ Col\ (Crypt\ c\ (Nonce\ Nv)) \in set\ evs6$;
  $Says\ Col\ Col\ (Nonce\ Nv) \notin set\ evs6$⟧
  $\implies Says\ Col\ Col\ (Nonce\ Nv)\ \#\ evs6 \in foo$

5 Proving Voter Privacy for FOO

5.1 Main Results

The following theorem, *foo_V_privacy_asynth*, is the culmination of the entire proof process and states that the FOO protocol guarantees voter privacy to all honest voters that started the protocol. More precisely, assume that the regular, honest voter V sent the administrator a message in line with the first step of the protocol, containing a blinded commitment on the vote Nv. Also assume that this very vote is in the message set of association syntheses. Then the name of V is not in that set:

theorem *foo_V_privacy_asynth*:
⟦$Says\ V\ Adm\ \{\!|Agent\ V,\ Crypt\ (priSK\ V)\ (Crypt\ b\ (Crypt\ c\ (Nonce\ Nv)))|\!\} \in set\ evs$;
  $a \in (asynth\ (aanalz\ Spy\ evs))$;
  $Nonce\ Nv \in a$; $V \notin bad$; $V \neq Adm$; $V \neq Col$; $evs \in foo$⟧
  $\implies Agent\ V \notin a$

Before turning to the proof itself, we focus on the most important proof elements, which are mainly results about associations.

A fundamental result is *foo_V_privacy_aanalz*, which looks similar to the *foo_V_privacy_asynth* theorem. However, whereas the latter is a statement about *asynth*, hence about association synthesis, the former only considers *aanalz*, that is associations arising from individual messages. Whenever an honest voter performed the first step of the protocol, the voter's identity and vote cannot be found in the same association:

theorem *foo_V_privacy_aanalz*:
⟦*Says V Adm* ⦃*Agent V*, *Crypt* (*priSK V*) (*Crypt b* (*Crypt c* (*Nonce Nv*)))⦄ ∈ *set evs*;
   a ∈ (*aanalz Spy evs*); *Nonce Nv* ∈ *a*; *V* ∉ *bad*; *evs* ∈ *foo*⟧
   ⟹ *Agent V* ∉ *a*

The lemma called *asynth_insert* is a direct consequence of the definition of *asynth* quoted in 3.2. By introducing the various cases that an application of *asynth* may imply, it provides a useful rewrite rule for expressions involving the operator name:

lemma *asynth_insert*:
a ∈ *asynth*(*insert a1 as*) ⟹
(*a=a1* ∨
a ∈ *asynth as* ∨
(∃ *a2 m*. *a2* ∈ *as* ∧ *a* = *a1* ∪ *a2* ∧ *m* ∈ *a1* ∧ *m* ∈ *a2* ∧
       m ≠ *Agent Adm* ∧ *m* ≠ *Agent Col*))

The next three theorems allow more precise reasoning about messages that contain encryption. They are all concerned with the situation where a message yields an association set containing at least one ciphertext. They are necessary for dealing with situations where protocol messages are not completely specified. For instance, an agent may have to transmit an encrypted commitment without even being able to check that the commitment is actually about a vote. In those situations, protocol step specification must model agents' limited knowledge when dealing with sealed messages. However, even when the complete contents of a ciphertext is not known, a number of scenarios can be distinguished. Various encryption key values and partial knowledge of the ciphertext contents lead to contradictions. Possible configurations are therefore made explicit in the following results.

Lemma *aanalz_PR* states constraints about the possible forms of a generic ciphertext appearing in any association. Its conclusion is expressed as a conjunction between two predicates that are themselves disjunctions. The first conjunct relates to the presence of an agent name in the association. If the name of the collector appears in the association and any nonce (a vote) is an atomic component of *R*, then no agents that are both honest and different from the collector can also be in *a*. The second conjunct states that if any nonce is part of the association, then the Spy must be able to decrypt the ciphertext and no agent name can be an atomic component of *R*:

lemma *aanalz_PR*:
⟦*a* ∈ *aanalz Spy evs*; *Crypt P R* ∈ *a*; *evs* ∈ *foo*⟧ ⟹
  (*Agent Col* ∉ *a* ∨
   (*Agent V* ∈ *a* ⟶ *V* ∈ *bad* ∨ *V* = *Col*) ∨
   (*Nonce Nv* ∉ *parts* {*R*})) ∧
  ((*Nonce Nv* ∉ *a*) ∨
   (*Key* (*invKey P*) ∈ *analz* (*spies evs*) ∧ *Agent V* ∉ *parts* {*R*}))

Then, lemma *aanalz_AdmPR_V_Nparts* relates to the specific case when a ciphertext signed by the administrator is in an association. It establishes a disjunction:

either no nonce is an atomic component of the ciphertext's body, or the Spy cannot open the ciphertext inside the signature, or there is no regular, honest agent name in the association:

lemma *aanalz_AdmPR_V_Nparts*:
$\llbracket a \in$ *aanalz Spy evs*; *Crypt* (*priSK Adm*) (*Crypt P R*) $\in a$; *evs* $\in$ *foo*$\rrbracket$
$\implies$ *Nonce Nv* $\notin$ *parts* $\{R\}$ $\vee$
  Key (*invKey P*) $\notin$ *analz* (*knows Spy evs*) $\vee$
  (*Agent V* $\in a \longrightarrow V \in$ *bad* $\vee V = Adm \vee V = Col$)

Finally, lemma *aanalz_Adm* is still about associations containing a ciphertext. Like *foo_V_privacy_aanalz*, it binds the variables involved in a version of the first step of the protocol. Assume an association contains the name of an honest agent who already sent a message corresponding to step one. Also assume it contains a ciphertext *Crypt P R*, and that the nonce from step one is in *parts* of *R*. If the name of the collector is absent from the association, then the following conclusions hold:

- If P is neither the signing key of the voter mentioned in the precondition nor the signing key of the administrator, then it must be the blinding factor;
- If P is the administrator's signing key, then the body of the ciphertext is exactly the body of the message signed by the voter in the bound first message:

lemma *aanalz_Adm*:
$\llbracket$*Says V Adm* $\{$*Agent V*, *Crypt* (*priSK V*) (*Crypt b* (*Crypt c* (*Nonce Nv*)))$\} \in$ *set evs*;
  $a \in$ *aanalz Spy evs*; *Agent Col* $\notin a$; *Agent V* $\in a$; $V \notin$ *bad*;
  Crypt P R $\in a$; *Nonce Nv* $\in$ *parts* $\{R\}$; *evs* $\in$ *foo*$\rrbracket$
$\implies (P = priSK V \vee P = priSK Adm \vee P = b) \wedge$
  $(P \neq priSK Adm \vee R = Crypt b (Crypt c (Nonce Nv)))$

5.2   Proof of the Main Theorem

Proving privacy by *foo_V_privacy_asynth* is done, as usual in the Inductive Method, by induction on the protocol model. Every protocol step generates a subgoal. When all subgoals are closed, the theorem is proven. Developing the proof required considerable effort. After eliminating redundancies and streamlining, it was reduced to about 170 steps. It will be shown that despite its length, the proving strategy is general, hence reusable for different protocols.

Induction and simplification leaves us with seven subgoals: the six protocols steps, plus *Fake*. The *Fake* is closed thanks to the classical reasoner *blast*. Its proof is simple because messages sent by dishonest agents do not yield associations. Intuitively, the goal of the Spy is to extract plausible associations, not make up new ones. However, it keeps its traditional Dolev-Yao attacker role and influences all usual theorems proven for the protocol ; those are used in the privacy proof.

The subgoal arising from *EV1* is first simplified by remarking that fresh keys (the blinding factor and commitment key) can never be known to the

Spy — they cannot yet be in the set *analz (knows Spy evs1)*. We then perform a case split about the agent *Va* involved in the version of the first protocol
step generated by this subgoal. If *Va* is dishonest (a member of the *bad* set),
then the message it sent yields no new association and the subgoal concludes
thanks to the inductive hypothesis. If *Va* is an honest agent, we must apply, for
the first time, *asynth_insert*. This lemma is of constant use throughout the proof
because it allows us to split the *asynth* set. For instance, this stage of the proof
features the following precondition:

a ∈ asynth (insert
{⦃Agent Va, Crypt (priSK Va) (Crypt ba (Crypt ca (Nonce Nva)))⦄,
Agent Va, Crypt (priSK Va) (Crypt ba (Crypt ca (Nonce Nva))),
Agent Va, Agent Adm, Crypt ba (Crypt ca (Nonce Nva))}
(aanalz Spy evs1))

Let us call X the set such that $a ∈ asynth$ (insert X (aanalz Spy evs1)).
  Applying *asynth_insert* leaves us with three possibilities:

1. $a = X$.
2. $a ∈ asynth$ (aanalz Spy evs1).
3. There exists *a2* in *aanalz Spy evs1* and an element *m* such that *a* is the
   union of *a2* and X and *m* is both in X and in *a2*.

The inductive hypothesis tells us that *Nv* is in *a* and X contains no nonces,
so the first disjunct is excluded. The second disjunct is eliminated thanks to
the lemma *nv_fresh_a2*, not quoted here, which states that fresh nonces do not
appear in association syntheses.

  If *a* is a union, more precision is required. First, if the agent *V* from the
inductive hypothesis and the agent *Va* introduced by the induction are different, then $V ∉ X$ and therefore *V* must be in *a2*. Since *Nv* is also in *a2*,
foo_V_privacy_aanalz leads us to a contradiction.

  Otherwise, $V = Va$. If *Nv* and *Nva* are equal, *Nv* must be fresh like *Nva*. The
auxiliary lemma *aanalz_traffic*, according to which elements in associations which
are not agent names must have appeared in the traffic, solves this case (fresh
elements never appeared in network traffic). On the other hand, if $Nv ≠ Nva$, the
element in common *m* can be any of the elements in X. We appeal to another
lemma, *association_Nv*. It is specifically tailored for this subgoal, used only here,
and shows that an association containing a nonce can not contain also any of the
possibilities for *m* listed here except for *V*. Together with *foo_V_privacy_aanalz*,
that takes care precisely of the case $m = V$, this solves the subgoal.

  The use of *asynth_insert* to split the association synthesis is a technique used
for all subgoals of the theorem. It turns out that the third disjunct generates the
bulk of the proving work for the remaining subgoals. We will therefore focus on
it. It requires taking a close look at the structure of sets in *aanalz*.

  Subgoals arising from protocol steps two and four are much larger than the
other ones because of the generic specification of the steps. For instance, in the
second protocol step, the administrator has received a signed ciphertext from
the voter. The administrator can extract the ciphertext from the signature, but

has no means in general to look inside. We only assume that it is possible to know that the ciphertext contains no more than one atomic component, by inspecting its length. However, the precise nature of the plaintext is unknown in general and this generality in the specification of the inductive step explains the additional complexity of the proof. It is necessary if the precondition is to be realistic. Likewise, in step four, the collector receives a signed ciphertext that he cannot open in general. The concrete consequence in terms of association syntheses is that potential common elements m are not listed explicitly in the goal preconditions. Instead of belonging to a finite set of bound variables, only partial information is known about them. For instance, we may only know that an element m can be deduced from some ciphertext via *analzplus*. By contrast, for non-generic protocol steps, we obtain an explicit set and the proof is much easier.

A number of results about elements in *aanalz* are available, such as *aanalz_PR* and *aanalz_Adm*. These theorems are stated with weak premises and offer a number of conclusions as disjunctions. The most systematic proof strategy is therefore to perform case splits about the ciphertext contained in *aanalz*. As this is done, one can reason more precisely about encryption keys and plaintexts until a contradiction is reached thanks to the aforementioned results. One crucial distinction is whether the name of the collector appears in the association. If such is the case, the elements in *aanalz* arose from the collection step *EV4*. Conversely, if *Agent Col $\notin$ a2*, the association set in the precondition was generated by another protocol step. The encryption key P from the ciphertext *Crypt P R*, assumed to be in an association, is then compared in turn to the voter's signing key, the administrator's signing key, and to the blinding factor. Contradictions are reached in every case. The value of the payload R is also compared with the voter's blinded vote commitment *Crypt b (Crypt c (Nonce Nv))*. Those different situations obviously refer to various ciphertext values naturally generated by the protocol steps. In essence, the proving strategy amounts to zooming in sufficiently into the various possible association configurations to uncover contradictions that are not apparent at a more general level.

The outline of this proving strategy is not dependent of a given protocol. Let us recall the important steps:

1. For every subgoal, split the association syntheses set *asynth* using *asynth_insert*.
2. The subgoals arising from explicit protocol steps are straightforward to close because the set of potential common elements m becomes explicit as well.
3. For more general subgoals, case splits about the possible values of initially generic ciphertexts are combined with lemmas describing their structure in associations in a systematic way.

5.3   Proof of the Supporting Theorems

Rather than describing the full proof of every theorem required for the privacy one, we focus on *aanalz_PR* due to space constraints. It is of constant use in the privacy proof, appearing in it eleven times, and its proof exemplifies the kind of reasoning required for the other supporting theorems. Recall its statement from earlier (5.1) ; it constrains the form of elements of *aanalz* that contain a ciphertext.

As expected, complications arise again from the generic steps, namely *EV2* and *EV4*. As the other subgoals are easier to prove, let us concentrate on *EV2*, as the proof for *EV4* is similar.

We require the following subsidiary results in addition to standard lemmas from the existing Inductive Method framework:

- *analzplus_into_parts*: Elements in the set *analzplus X ks* (recall *ks* is the external key set) are in *parts X*.
- *no_pairs*: If a message contains only atomic components and already contains an agent name in its *parts* set, a number of other elements cannot be in the *parts* set.
- *analzplus_Nv*: Assume an *analzplus Q (analz H)* set contains a ciphertext and a nonce. If *parts Q* contains only atomic components, then the decryption key of the ciphertext must be in *analz (insert Q H)*.

The case where the administrator (*Adm*) is dishonest is closed easily. Else, we must distinguish cases on the basis of the origin of the association in the inductive hypothesis. The first possibility is that it was generated by the *EV2* message introduced by the induction. In that case, the key *P* in the *aanalz_PR* theorem statement could either be the administrator's signing key, or the encryption key of the ciphertext that he signed. A third possibility is that the entire *Crypt P R* is embedded deeper in the signed ciphertext. Let us examine each possibility in turn.

In the first case, we must show that an agent name (either the collector or a regular voter) and a nonce cannot be present in *analzplus R* at the same time. This is shown by combining *analzplus_into_parts* and *no_pairs*. In the second case, the ciphertext *Crypt P R* from the theorem precondition is exactly the ciphertext *Crypt Pa Ra* signed by the administrator in this version of the second protocol step. Disentangling the precondition conjunction leads to the same scenario and an additional one that entails proving that if the inverse of key *P* is known to the attacker in the first place, it is all the more known to her after getting hold of *R*.

If *Crypt P R* is embedded in the ciphertext generated by the administrator, we must perform a few additional case splits but the line of reasoning is the same, with the additional use of *analzplus_Nv*.

Even though specifying the possible forms of elements in *aanalz* requires inspecting a number of scenarios, the proving process is straightforward once some crucial building blocks are established. Notably, the three subsidiary results we listed earlier (*analzplus_into_parts* and so on) are stated without any reference to the FOO protocol — they are protocol-independent and can be reused directly. A submission of our Isabelle theories to the online Archive of Formal Proofs [1] is being prepared.

6   Conclusion

We have presented the first interactive theorem proving-based analysis of voter privacy. It offers an independent and complementary means of investigation to consolidated work based on process equivalence, ultimately contributing to the trustworthiness of voting systems. Privacy is modelled as an unlinkability property between a voter and her ballot. Extensions to the Inductive Method are implemented in Isabelle/HOL to specify associations between elements and combinations of associations that share a common element.

The initial proof development effort was significant, but a coherent line of reasoning emerges from the proof. This general strategy and a number of protocol-independent results about the new operators support the case of re-usability for other e-voting protocols. Interactive proofs entail a level of clarity about protocol scenarios that is unavailable from automatic tools. The inductive nature of our specification eliminates termination issues or inherent size limitations. While the benefits of automated tools are clear, our approach sheds a complementary light on voter privacy by its operational view.

A more general version of the *asynth* operator, allowing unbounded association synthesis, is needed. Other privacy-type properties such as receipt-freeness and coercion-resistance ought to be specified in the Inductive Method. Additionally, e-voting protocols that are not amenable to analysis in the process equivalence model must be studied in our framework to investigate its domain of applicability. We would also like to program some of the recurring proof steps as ML tactics.

Acknowledgement. This research was supported in part by the Science Foundation Ireland (SFI) grant 08/RFP/CMS1347.

References

1. Klein, G., Nipkow, T., Paulson, L. (eds.): The Archive of Formal Proofs, http://afp.sf.net
2. Abadi, M., Fournet, C.: Mobile Values, New Names, and Secure Communication. In: Proc. of the 28th ACM SIGACT-SIGPLAN Symposium on Principles of Programming Languages (POPL 2001), pp. 104–115. ACM Press (2001)
3. Bella, G.: Formal Correctness of Security Protocols. Information Security and Cryptography. Springer (2007)
4. Bella, G.: Inductive study of confidentiality: for everyone. Formal Aspects of Computing, 1–34 (2012)
5. Bella, G., Massacci, F., Paulson, L.C., Tramontano, P.: Formal Verification of Cardholder Registration in SET. In: Cuppens, F., Deswarte, Y., Gollmann, D., Waidner, M. (eds.) ESORICS 2000. LNCS, vol. 1895, pp. 159–174. Springer, Heidelberg (2000)
6. Bella, G., Paulson, L.C.: Mechanical Proofs about a Non-Repudiation Protocol. In: Boulton, R.J., Jackson, P.B. (eds.) TPHOLs 2001. LNCS, vol. 2152, pp. 91–104. Springer, Heidelberg (2001)

7. Blanchet, B.: An Efficient Cryptographic Protocol Verifier Based on Prolog Rules. In: Proc. of the 14th IEEE Computer Security Foundations Workshop (CSFW 2001), pp. 82–96. IEEE Press (1998)
8. Butin, D., Bella, G.: Verifying Privacy by Little Interaction and No Process Equivalence. In: SECRYPT, pp. 251–256. SciTePress (2012)
9. Chadha, R., Ciobâcă, Ş., Kremer, S.: Automated Verification of Equivalence Properties of Cryptographic Protocols. In: Seidl, H. (ed.) ESOP 2012. LNCS, vol. 7211, pp. 108–127. Springer, Heidelberg (2012)
10. Delaune, S., Kremer, S., Ryan, M.: Verifying privacy-type properties of electronic voting protocols. Journal of Computer Security 17(4), 435–487 (2009)
11. Delaune, S., Ryan, M., Smyth, B.: Automatic Verification of Privacy Properties in the Applied pi Calculus. In: Karabulut, Y., Mitchell, J., Herrmann, P., Jensen, C.D. (eds.) Trust Management II. IFIP, vol. 263, pp. 263–278. Springer, Boston (2008)
12. Fujioka, A., Okamoto, T., Ohta, K.: A Practical Secret Voting Scheme for Large Scale Elections. In: Zheng, Y., Seberry, J. (eds.) AUSCRYPT 1992. LNCS, vol. 718, pp. 244–251. Springer, Heidelberg (1993)
13. Kremer, S., Ryan, M.: Analysis of an Electronic Voting Protocol in the Applied Pi Calculus. In: Sagiv, M. (ed.) ESOP 2005. LNCS, vol. 3444, pp. 186–200. Springer, Heidelberg (2005)
14. Martina, J.E., Paulson, L.C.: Verifying Multicast-Based Security Protocols Using the Inductive Method. In: Workshop on Formal Methods and Cryptography (CryptoForma 2011) (2011)
15. Paulson, L.C.: Isabelle. LNCS, vol. 828. Springer, Heidelberg (1994)
16. Paulson, L.C.: The Inductive Approach to Verifying Cryptographic Protocols. Journal of Computer Security 6, 85–128 (1998)
17. Ryan, M.: Keynote: Analysing security properties of electronic voting systems. In: Erlingsson, Ú., Wieringa, R., Zannone, N. (eds.) ESSoS. LNCS, vol. 6542, pp. 1–14. Springer (2011)

Confidentiality for Probabilistic Multi-threaded Programs and Its Verification

Tri Minh Ngo, Mariëlle Stoelinga, and Marieke Huisman

University of Twente, Netherlands
tringominh@gmail.com,
{Marielle.Stoelinga,Marieke.Huisman}@ewi.utwente.nl

Abstract. Confidentiality is an important concern in today's information society: electronic payment and personal data should be protected appropriately. This holds in particular for multi-threaded applications, which are generally seen the future of high-performance computing. Multi-threading poses new challenges to data protection, in particular, data races may be exploited in security attacks. Also, the role of the scheduler is seminal in the multi-threaded context.

This paper proposes a new notion of confidentiality for probabilistic and non-probabilistic multi-threaded programs, formalized as scheduler-specific probabilistic observational determinism (SSPOD), together with verification methods. Essentially, SSPOD ensures that no information about the private data can be derived either from the public data, or from the probabilities of the public data being changed. Moreover, SSPOD explicitly depends on a given (class of) schedulers.

Formally, this is expressed by using two conditions: (i) each publicly visible variable individually behaves deterministically with probability 1, and (ii) for every trace considering all publicly visible variables, there always exists a matching trace with equal probability. We verify these conditions by a clever combination of new and existing algorithms over probabilistic Kripke structures.

1 Introduction

Confidentiality plays a crucial role in the development of applications dealing with private data, such as Internet banking, medical information systems, and authentication systems. These systems need to enforce strict protection of private data, like credit card details, medical records, etc. The key idea is that secret information should not be derivable from public data. For example, the program if $(h > 0)$ then $l := 0$ else $l := 1$, where h is a private variable and l is a public variable[1], is considered insecure, because we can derive the value of h from the value of l. If private data is not sufficiently protected, users refuse to use such applications. Using formal means to establish confidentiality is a promising way to gain the trust of users.

---

[1] For simplicity, throughout this paper, we consider a simple two-point security lattice, where the data is divided into two disjoint subsets, of private (high) and public (low) security levels, respectively.

J. Jürjens, B. Livshits, and R. Scandariato (Eds.): ESSoS 2013, LNCS 7781, pp. 107–122, 2013.

Possibilistic programs. With the trend of multiple cores on a chip and massively parallel systems like general purpose graphic processing units, multi-threading is becoming more standard. Existing confidentiality properties, such as *noninterference* [12] and *observational determinism* [31,15] are not suitable to ensure confidentiality for multi-threaded programs. They only consider input-output behavior, and ignore the role of schedulers, while multi-threaded programs allow all interactions between threads and intermediate results to be observed [31,15,14]. Thus, new methods have to be developed for an observational model where an attacker can access the full code of the program, observe the traces of public data, and limit the set of possible program traces by selecting a scheduler.

Because of the exchange of intermediate results, to ensure confidentiality for multi-threaded programs, it is necessary to consider the whole execution traces, i.e., the sequences of states that occur during the execution of the program [31,24]. Besides, due to the interactions between threads, the traces of a multi-threaded program depend on the scheduler that is used to execute the program. Therefore, a program's confidentiality is only guaranteed under a particular scheduler, while a different scheduler might make the program reveal secret information, as illustrated by the following example.

$$\{\text{if } (\text{h} > 0) \text{ then } \text{l1} := 1 \text{ else } \text{l2} := 1\} \| \{\text{l1} := 1; \text{l2} := 1\} \| \{\text{l2} := 1; \text{l1} := 1\},$$

where $\|$ is the parallel operator. Under a nondeterministic scheduler, the secret information cannot be derived, because the traces in the cases $\text{h} > 0$ and $\text{h} \leq 0$ are the same. However, under a scheduler that always executes the leftmost thread first, the secret information is revealed by observing whether l1 is updated before l2, i.e., when l1 is updated before l2, the attacker knows that $\text{h} > 0$. However, this program is considered secure by observational determinism [31,15].

Taking into account the effect of schedulers on confidentiality, we proposed a definition of *scheduler-specific observational determinism* (SSOD) for *possibilistic* multi-threaded programs [14]. Basically, a program respects SSOD if (SSOD-1) for any initial state, traces of each public variable are stuttering-equivalent, and (SSOD-2) for any two initial states I and I' that are indistinguishable w.r.t. the public variables, for every trace starting in I, there *exists* a trace that is stuttering equivalent w.r.t. *all* public variables, starting in I'.

SSOD is scheduler-specific, since traces model the runs of a program under a particular scheduler. When the scheduling policy changes, some traces *cannot occur*, and also, some new traces *might appear*; thus the new set of traces may not respect our requirements. For example, the above program is accepted by SSOD w.r.t. the nondeterministic scheduler, but is rejected under the scheduler that always executes the leftmost thread first.

Probabilistic programs. To extend our earlier results, this paper also considers programs that have probabilistic behaviors. For probabilistic programs, some threads might be more likely to be executed than others. This opens up the possibility of probabilistic attacks, as in the following example.

if (h > 0) then {l1 := 1 || l2 := 1} else {l1 := 1 || l2 := 1}.

This program is secure under a nondeterministic scheduler. However, consider a scheduler that, when h > 0, picks thread l1:=1 first with probability 3/4; otherwise, it chooses between the threads with equal probabilities. With this scheduler, we can learn information about h from the probabilities of public data traces. However, the program is still accepted by SSOD w.r.t. this scheduler, because SSOD only considers the existence of traces, not its probability.

To detect vulnerabilities to probabilistic attacks, we define *scheduler-specific probabilistic-observational determinism* (SSPOD). This formalizes the observational determinism property for probabilistic multi-threaded programs, executed under a probabilistic scheduler. Basically, a program respects SSPOD if (SSPOD-1) for any initial state, each public variable individually behaves deterministically with probability 1, and (SSPOD-2) for any two initial states I and I' that are indistinguishable *w.r.t.* the public variables, for every trace starting in I, there *exists* a trace that is stuttering equivalent *w.r.t. all* public variables, starting in I', and the probabilities of these two matching traces are the same.

The first condition of SSPOD requires that all public variable traces individually evolve deterministically. Requiring only that a stuttering-equivalent public variable trace exists is not sufficient to guarantee confidentiality for multi-threaded programs, as extensively discussed in [31,14]. The first condition avoids leakage of private information based on the observation of public data traces. The second condition of SSPOD requires the existence of a public data trace with equal probabilities. This existential condition avoids refinement attacks where an attacker chooses an appropriate scheduler to control the set of possible traces. The second condition is also sufficient to ensure that any difference in the relative order of updates is coincidental, and thus no private information can be deduced from it. In addition, SSPOD also guarantees that no private information can be derived from the probabilistic distribution of traces, because indistinguishable traces occur with the same probabilities.

Notice that the question which classes of schedulers appropriately model real-life attacks is orthogonal to our results: our definition is parametric on the scheduler. In Section 5, we compare SSPOD with the existing formalizations of confidentiality properties for probabilistic programs [30,24,25], and argue that they are either unsuitable to the multi-threaded context, or very restrictive.

Verification. Besides formalizing the property, the paper also discusses how to verify SSPOD. The traditional way to check information flow properties is by using a type system. However, as discussed in [14], type systems are not suited to verify existential properties, as the one in SSPOD. Besides, type systems that have been proposed to enforce confidentiality for multi-threaded programs are often very restrictive. This restrictiveness makes the application programming become impractical; many intuitively secure programs are rejected by this approach, i.e., h := 1; 1 := h. Instead, in [14], we proposed to use a different approach for SSOD, encoding the information flow property as a temporal logic property. This idea is based on the use of self-composition [6,15], and allows us to verify the information flow property via model checking. However, the result is rather complex, and thus its verification cannot be handled efficiently by the existing model-checking tools.

Therefore, this paper proposes more efficient algorithms to verify our definition. For this purpose, programs are modeled as probabilistic Kripke structures. For both conditions of SSPOD, we present a verification approach, a clever combination of new and existing algorithms. The first condition is checked by removing all stuttering loops, except the self-loops in final states, and then verifying stuttering equivalence. Verification of stuttering equivalence is implemented by checking whether there exists a functional bisimulation between the executions of the Kripke structure and a witness trace. This is a new algorithm, that is also relevant outside the security context, e.g., as in partial-order reduction for model checking, because stuttering equivalence is a fundamental concept in the theory of concurrent and distributed systems. SSOD-1 can be also verified by a variant of this algorithm. SSPOD-2 is implemented by removing stuttering steps, thereby reducing the problem into an equivalence problem for probabilistic languages [29,11,16]. This approach gives a precise verification method for observational determinism. Furthermore, the model checking procedure is also able to produce a counter-example to synthesize attacks for insecure programs, i.e., for programs that fail either of the conditions of SSPOD (similar as in [21]).

Currently, we are implementing our verification techniques in the symbolic model checker LTSmin [7]. SSPOD-1 has been implemented, and we will adapt the existing implementation of [16] for SSPOD-2. Once the implementation is finished, we will apply the tool to case studies.

Organization of the Paper. Section 2 presents the preliminaries. Then, Section 3 formalizes the SSPOD property, and Section 4 presents its verification. Section 5 discusses related work. Section 6 concludes, and discusses future work.

2   Preliminaries

2.1   Basics

Sequences. Let X be an arbitrary set. The sets of all *finite sequences*, and all *sequences* of X are denoted by X^*, and X^ω, respectively. The empty sequence is denoted by ε. Given a sequence $\sigma \in X^*$, we denote its last element by $last(\sigma)$. A sequence $\rho \in X^*$ is called a *prefix* of σ, denoted by $\rho \sqsubseteq \sigma$, if there exists another sequence $\rho' \in X^\omega$ such that $\rho\rho' = \sigma$.

Probability distributions. A *probability distribution* μ over a set X is a function $\mu \in X \to [0,1]$, such that the sum of the probabilities of all elements is 1, i.e., $\sum_{x \in X} \mu(x) = 1$ over a set X. If X is uncountable, then $\sum_{x \in X} \mu(x) = 1$ implies that $\mu(x) > 0$ for countably many $x \in X$. We denote by $\mathcal{D}(X)$ the set of all probability distributions over X. The *support* of a distribution $\mu \in \mathcal{D}(X)$ is the set $supp(\mu) = \{x \in X \mid \mu(x) > 0\}$ of all elements with a positive probability. For an element $x \in X$, we denote by $\mathbf{1}_x$ the probability distribution that assigns probability 1 to x and 0 to all other elements.

2.2 Probabilistic Kripke Structures

We consider probabilistic Kripke structures (PKS) that can be used to model semantics of probabilistic programs in a standard way [13]. PKSs are like standard Kripke structures [17], except that each transition $c \rightarrow \mu$ leads to a probability distribution μ over the next states, i.e., the probability to end up in state c' is $\mu(c')$. Each state may enable several probabilistic transitions, modeling different execution orders to be determined by a scheduler. For technical convenience, our PKSs label states with arbitrary-valued variables from a set *Var*, rather than with Boolean-valued atomic propositions. Thus, each state c is labeled by a labeling function $V(c) :$ *Var* $\rightarrow$ *Val* that assigns a value $V(c)(v) \in$ *Val* to each variable $v \in$ *Var*. We assume that *Var* is partitioned into sets of low variables L and high variables H, i.e., $Var = L \cup H$, with $L \cap H = \emptyset$.

Definition 1 (Probabilistic Kripke structure). *A probabilistic Kripke structure $\mathcal{A}$ is a tuple $\langle S, I, Var, Val, V, \rightarrow \rangle$ consisting of (i) a set S of states, (ii) an initial state $I \in S$, (iii) a finite set of variables Var, (iv) a countable set of values Val, (v) a labeling function $V :$ $S \rightarrow$ (Var $\rightarrow$ Val), (vi) a transition relation $\rightarrow \subseteq S \times \mathcal{D}(S)$. We assume that $\rightarrow$ is non-blocking, i.e., $\forall c \in S. \exists \mu \in \mathcal{D}(S). c \rightarrow \mu$.*

A PKS is *fully probabilistic* if each state has at most one outgoing transition, i.e., if $c \rightarrow \mu$ and $c \rightarrow \mu'$ implies $\mu = \mu'$. Given a set $Var' \subseteq Var$, the *projection* $\mathcal{A}_{|Var'}$ of $\mathcal{A}$ on Var', restricts the labeling function V to labels in Var'. Thus, we obtain $\mathcal{A}_{|Var'}$ from $\mathcal{A}$ by replacing V by $V_{\cdot|Var'} :$ $S \rightarrow (Var' \rightarrow Val)$.

Semantics of probabilistic programs. A program C over a variable set *Var* can be expressed as a PKS $\mathcal{A}$ in a standard way: The states of $\mathcal{A}$ are tuples $\langle C, s \rangle$ consisting of a program fragment C and a valuation $s :$ *Var* $\rightarrow$ *Val*. The transition relation $\rightarrow$ follows the small-step semantics of C. If a program terminates in a state c, we include a special transition $c \rightarrow \mathbf{1}_c$, ensuring that $\mathcal{A}$ is non-blocking.

Paths and traces. A *path* π in $\mathcal{A}$ is an infinite sequence $\pi = c_0 c_1 c_2 \ldots$ such that (i) $c_i \in S, c_0 = I$, and (ii) for all $i \in \mathbb{N}$, there exists a transition $c_i \rightarrow \mu$ with $\mu(c_{i+1}) > 0$. We define $Path(\mathcal{A})$ as the set of all infinite paths of $\mathcal{A}$; and $Path^*(\mathcal{A}) = \{\pi' \sqsubseteq \pi \mid \pi \in Path(\mathcal{A})\}$ as the set of all finite paths in $Path(\mathcal{A})$.

The *trace* T of a path π records the valuations along π. Formally, $T = trace(\pi) = V(c_0)V(c_1)V(c_2)\ldots$. Trace T is a *lasso* iff it ends in a loop, i.e., if $T = T_0 \ldots T_i (T_{i+1} \ldots T_n)^\omega$, where $(T_{i+1} \ldots T_n)^\omega$ denotes a loop. Let $Trace(\mathcal{A})$ denote the set of all infinite traces of $\mathcal{A}$. Two states c and c' are *low-equivalent*, denoted $c =_L c'$, iff $V(c)_{|L} = V(c')_{|L}$.

2.3 Probabilistic Schedulers

A probabilistic scheduler is a function that implements a scheduling policy [24], i.e., that decides with which probabilities the threads are selected. To make our security property applicable for many schedulers, we give a general definition. We allow a scheduler to use the full history of computation to make decisions: given a path ending in some state c, a scheduler chooses which of the probabilistic transitions enabled in c to execute. Since each transition results in a distribution, a probabilistic scheduler returns a distribution of distributions[2].

Definition 2. *A scheduler δ for PKS $\mathcal{A} = \langle S, I, Var, Val, V, \to \rangle$ is a function $\delta : Path^*(\mathcal{A}) \to \mathcal{D}(\mathcal{D}(S))$, such that, for all finite paths $\pi \in Path^*(\mathcal{A})$, $\delta(\pi)(\mu) > 0$ implies $last(\pi) \to \mu$.*

The effect of a scheduler δ on a PKS $\mathcal{A}$ can be described by a PKS $\mathcal{A}_\delta$: the set of states of $\mathcal{A}_\delta$ is obtained by unrolling the paths in $\mathcal{A}$, i.e., $S_{\mathcal{A}_\delta} = Path^*(\mathcal{A})$ such that states of $\mathcal{A}_\delta$ contain a full history of execution. Besides, the unreachable states of $\mathcal{A}$ under the scheduler δ are removed by the transition relation $\to_\delta$.

Definition 3. *Let $\mathcal{A} = \langle S, I, Var, Val, V, \to \rangle$ be a PKS and let δ be a scheduler for $\mathcal{A}$. The PKS associated to δ is $\mathcal{A}_\delta = \langle Path^*(\mathcal{A}), I, Var, Val, V_\delta, \to_\delta \rangle$, where $V_\delta : Path^*(\mathcal{A}) \times Var \to Val$ is given by $V_\delta(\pi) = V(last(\pi))$, and the transition relation is given by $\pi \to_\delta \mu$ iff $\mu(\pi c) = \sum_{\nu \in supp(\delta(\pi))} \delta(\pi)(\nu) \cdot \nu(c)$ for all π, c.*

Since all nondeterministic choices in $\mathcal{A}$ have been resolved by δ, $\mathcal{A}_\delta$ is fully probabilistic. The probability $P(\pi)$ given to a finite path $\pi = \pi_0 \pi_1 \ldots \pi_n$ is determined by $\delta(\pi_0)(\pi_1) \cdot \delta(\pi_0 \pi_1)(\pi_2) \cdots \delta(\pi_0 \pi_1 \ldots \pi_{n-1})(\pi_n)$. The probability of a finite trace T is obtained by adding the probabilities of all paths associated with T. Based on this observation, we can associate a probability space $(\Omega, \mathcal{F}, \mathbf{P}_\delta)$ over sets of traces. Following the standard definition, we set $\Omega = (Var \to Val)^\omega$, $\mathcal{F}$ contains all measurable sets of traces, and $\mathbf{P}_\delta : \mathcal{F} \to [0, 1]$ is a probability measure on $\mathcal{F}$. Thus, $\mathbf{P}_\delta(X)$ is the probability that a trace inside set $X \in \mathcal{F}$ occurs. We refer to [27] for technical details. Notice that Ω and $\mathcal{F}$ depend only on $\mathcal{A}$, not on $\mathcal{A}_\delta$.

2.4 Stuttering-Free PKSs and Stuttering Equivalence

Stuttering steps and *stuttering equivalence* [22,15] are the basic ingredients of our confidentiality properties. In the non-probabilistic case, a stuttering step is a transition $c \to c'$ that leaves the labels unchanged, i.e., $V(c') = V(c)$. In the probabilistic scenario, a transition stutters if, with positive probability, at least one of the reached states has the same label. A *stuttering-free* PKS allows stuttering transitions only as the self-loops in final states.

---

[2] Thus, we assume a discrete probability distribution over the uncountable set $\mathcal{D}(S)$; only the countably many transitions occurring in $\mathcal{A}$ can be scheduled with a positive probability.

Definition 4 (Stuttering-free PKS). *A stuttering step is a transition $c \to \mu$ with $V(c) = V(c')$ for some $c' \in supp(\mu)$: A PKS is called stuttering-free if for all stuttering steps $c \to \mu$, we have that $\mu = \mathbf{1}_c$ and no other transition leaving from c, i.e., if $c \to \mu'$, this implies $\mu = \mu'$.*

Two sequences are stuttering equivalent if they are the same after we remove adjacent occurrences of the same label, e.g., $(\mathbf{aaabcccd})^{\omega}$ and $(\mathbf{abbcddd})^{\omega}$.

Definition 5 (Stuttering equivalence). *Let X be a set. Stuttering equivalence, denoted $\sim$, is the largest equivalence relation over $X^{\omega} \times X^{\omega}$ such that for all $T, T' \in X^{\omega}, a, b \in X: aT \sim bT' \Rightarrow a = b \wedge (T \sim T' \vee aT \sim T' \vee T \sim bT')$. A set $Y \subseteq X$ is* closed under stuttering equivalence *if $T \in Y \wedge T \sim T'$ imply $T' \in Y$.*

3   Scheduler-Specific Probabilistic-Observational Determinism

A program is confidential w.r.t. a particular scheduler iff no secret information can be derived from the observation of public data traces, the order of public data updates, or from the probabilities of traces. This is captured formally by the definition of *scheduler-specific probabilistic-observational determinism.*

As shown in [31,14], to be secure, a multi-threaded program must enforce an order on the accesses to a single low variable, i.e., the sequence of operations performed at a single low variable is deterministic. Therefore, SSPOD's first condition requires that for any initial state, traces of each low variable that do not end in a non-final stuttering loop are stuttering equivalent with probability 1. This condition ensures that no secret information can be derived from the observation of public data traces, because when all low variables individually evolve deterministically, the values of low variables do not depend on the values of high variables. However, a consequence of SSPOD's first condition is that harmless programs such as `1:=0 || 1:=1` are also rejected.

SSPOD also requires that, given any two initial low-equivalent states I and I', for every trace starting in I, there *exists* a trace that is stuttering equivalent *w.r.t. all* low variables, starting in I', and the probabilities of these two matching traces are the same. This condition ensures that secret information cannot be derived from the relative order of updates of any two low variables, or from any probabilistic attack, because there is always a matching trace with the same probability of occurrence.

Let $(\Omega, \mathcal{F}, \mathbf{P}_{\delta})$ denote the probability space of $\mathcal{A}_{\delta}$ with an initial state I. Notice that the probability of a trace to end up in a non-final stuttering loop is 0, because a non-final loop must contain at least one state with a transition that goes out of the loop; thus, it contains a transition with a probability less than 1. Thus, if X is a set of traces that ends in a non-final stuttering loop and are closed under stuttering equivalence, $\mathbf{P}_{\delta}[X]$ might be 0. Therefore, SSPOD is formally defined as follows.

Definition 6 (SSPOD). *Given a scheduler δ, a program C respects SSPOD w.r.t. L and δ, iff for any initial state I,*

SSPOD-1 *For any $l \in L$, let $X \in \mathcal{F}$ be any set of traces closed under stuttering equivalence w.r.t. l, we have $\mathbf{P}_\delta[X] = 1$ or $\mathbf{P}_\delta[X] = 0$.*

SSPOD-2 *For any initial state I' that is low-equivalent with I, for all sets of traces $X \in \mathcal{F}$ that are closed under stuttering equivalence w.r.t. L, we have $\mathbf{P}_\delta[X] = \mathbf{P}'_\delta[X]$, where $(\Omega, \mathcal{F}, \mathbf{P}'_\delta)$ denote the probability space of $\mathcal{A}_\delta$ with I'.*

Program C is scheduler-specific probabilistic-observational deterministic w.r.t. a set of schedulers Δ if it is so w.r.t. any scheduler $\delta \in \Delta$.

4   Verification of SSPOD

This section discusses how we algorithmically verify the two conditions of SS-POD. As mentioned above, we use a combination of new and existing algorithms. Moreover, the new algorithm is general, and also applicable in other, non-security related contexts. We assume that data domains are finite and schedulers use finite memory. Therefore, the algorithms work only on finite fully probabilistic PKSs, which can be viewed as finite Markov Chains.

4.1   Verification of SSPOD-1

Algorithm. Given a program C, and a scheduler δ, SSPOD-1 requires that after projecting $\mathcal{A}_\delta$ on any low variable l, all traces that do not stutter forever in a non-final stuttering loop must be stuttering equivalent with probability 1. To verify this, we pick one arbitrary trace and ensure that all other traces are stuttering equivalent to this trace. Concretely, for each $l \in L$, we carry out the following steps.

--------SSPOD-1 on l--
  1: Project $\mathcal{A}_\delta$ on l, yielding $\mathcal{A}_{\delta\,|_l}$.
  2: Remove all stuttering loops in $\mathcal{A}_{\delta\,|_l}$.
  3: Re-establish the self-loops for final states of $\mathcal{A}_{\delta\,|_l}$. This yields a stuttering-loop free PKS, denoted $\mathcal{R}_{\delta\,|_l}$.
  4: Check whether all traces of $\mathcal{R}_{\delta\,|_l}$ are stuttering equivalent by:
    4.1: Choose a *witness* trace by:
        4.1.1: Take an arbitrary lasso T of $\mathcal{R}_{\delta\,|_l}$.
        4.1.2: Remove stuttering steps and minimize T.
    4.2: Check stuttering trace equivalence between $\mathcal{R}_{\delta\,|_l}$ and T by checking if there exists a functional bisimulation between them.
_____

This algorithm works, since we transform the probabilistic property SSPOD-1 into a possibilistic one. Key insight is that the probability of a trace that stutters forever in a non-final stuttering loop is 0. Therefore, after removing all non-final stuttering loops, it is sufficient to determine whether all traces are stuttering equivalent.

To perform Step 1, we label every state with the value of l in that state. To remove the stuttering loops in Step 2, we use a classical algorithm for finding strongly connected components w.r.t. stuttering steps [1], and collapse these components into a single state. To ensure that the transition relation remains non-blocking, Step 3 re-establishes the self-loops for final states.

Step 4.1.1 is implemented via a classical cycle-detection algorithm based on depth-first search (Appendix A of [20]). The initial state of a lasso is also the initial state of PKS. The algorithm essentially proceeds by picking arbitrary next steps, and terminates when it hits a state that was picked before. Step 4.1.2 is done via the standard strong bisimulation reduction. For example, the minimal form of a lasso $\mathbf{abb(cb)}^\omega$ is $\mathbf{a(bc)}^\omega$. This minimal lasso is called the *witness trace*.

Step 4.2 checks stuttering trace equivalence between a PKS $\mathcal{A}$ and the witness trace T by checking if there exists a functional bisimulation between them, i.e., a bisimulation that is a function, thus mapping each state in $\mathcal{A}$ to a single state in T. This is done by exploring the state space of $\mathcal{A}$ in a breadth-first search (BFS) order and building the mapping *Map* during exploration. We name each state in T by a unique symbol $u \in \mathcal{U}$, i.e., u_i denotes T_i. Let $succ(T, u)$ denote the successor of u on T.

We map the $\mathcal{A}$'s initial state to u_0, i.e., $Map[init\_state] = u_0$. Each iteration of the algorithm examines the successors of the state stored in the variable *current*. Assume that $Map[current]$ is u, consider a successor $c \in succ(\mathcal{A}, current)$. The *potential_map* of c is u if $current \to c$ is a stuttering transition; otherwise, it is $succ(T, u)$. The algorithm returns *false*, i.e., $continue = false$, if (i) c and *potential_map* have different valuations, (ii) c is a final state of $\mathcal{A}$, while *potential_map* is not the final state of T, or (iii) c has been checked before, but its mapped state is not *potential_map*.

If none of these cases occurs and c was not checked before, c is added to Q, and mapped to *potential_map*. Basically, a state c of $\mathcal{A}$ is mapped to u, i.e., $Map[c] = u$, iff the trace from the initial state to state c in $\mathcal{A}$ and the prefix of T upto u are stuttering equivalent.

Let $c \sim_V c'$ denote that c and c' have the same valuation, i.e., $V(c) = V(c')$; $final(\mathcal{A}, c)$ denote that c is a final state in $\mathcal{A}$; and $final(T, u)$ denote that u is the final state in T. The algorithm also uses a FIFO queue Q of *frontier* states. The termination of Algorithm 4.2 follows from the termination of BFS over a finite $\mathcal{A}$.

4.2: Stuttering Trace Equivalence($\mathcal{A}$, T)

    for all states $c \in \mathcal{S}$ **do** $Map[c] := \bot$;
    $continue := true$;
    $Q := empty\_queue()$; $enqueue(Q, init\_state)$;
    $Map[init\_state] := u_0$; $// \; u_0 \text{ is } T_0$
    while $!empty(Q) \land continue$ **do**
        $current := dequeue(Q)$;
        $u := Map[current]$;
        for all states $c \in succ(\mathcal{A}, current)$ **do**

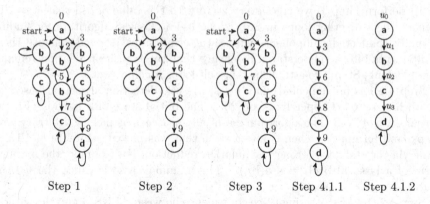

Step 1   Step 2   Step 3   Step 4.1.1   Step 4.1.2

Fig. 1. Step 1 - Step 4.1

$$potential\_map := (c \sim_V current) ? u : succ(T, u);$$
case $c \not\sim_V potential\_map \rightarrow continue := false;$
∥ $final(\mathcal{A}, c) \wedge \neg final(T, potential\_map) \rightarrow continue := false;$
∥ $Map[c] = \bot \rightarrow enqueue(Q, c);$
$$Map[c] := potential\_map;$$
∥ $Map[c] \neq potential\_map \rightarrow continue := false;$
return $continue;$

Example 1. Figure 1 illustrates Step 1 - Step 4.1 on a PKS $\mathcal{A}$ consisting of 10 states. Step 1 projects $\mathcal{A}$ on a low variable l. The symbols **a**, **b**, **c** etc. denote state contents, i.e., states with the same value of l are represented by the same symbol. Step 2 removes all stuttering loops, while Step 3 re-establishes the self-loops for final states. Step 4.1 takes an arbitrary trace of $\mathcal{A}$ and then minimizes it. Each state of the witness trace T is denoted by a unique symbol u_i. Figure 2 illustrates Step 4.2. Initially, all states of $\mathcal{A}$ are mapped to a special symbol $\bot$ that indicates unchecked states. To keep states readable, we skip the valuation. Next, state 0 is enqueued, and mapped to u_0. Next, the algorithm examines all unchecked successors of state 0, i.e., states 1, 2, 3. Each of them follows a non-stuttering step, thus their *potential_maps* are all u_1. Since states 1, 2, and 3 have the same valuation as *potential_map*, i.e., **b**, they are all enqueued, and mapped to u_1. Next, the successor of state 1, i.e., state 4, is considered. The transition $1 \rightarrow 4$ is non-stuttering, thus *potential_map* $= u_2$. State 4 has the same valuation as *potential_map*, but it is a final state of $\mathcal{A}$, while *potential_map* is not the final state of T. Thus, *continue* $= false$. The PKS $\mathcal{A}$ and the witness trace T are not stuttering trace equivalent, because there exists a trace that stutters in state 4 forever. The algorithm terminates.

Theorem 1. *Algorithm 4.2 returns true iff there exists a bisimulation between* $\mathcal{A}$ *and* T.

Proof. See Appendix B of [20].

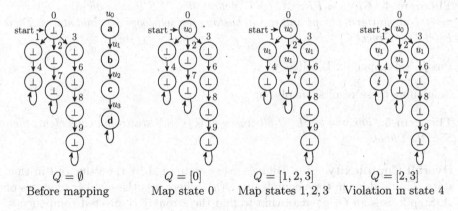

$Q = \emptyset$ $Q = [0]$ $Q = [1, 2, 3]$ $Q = [2, 3]$

Before mapping   Map state 0   Map states 1, 2, 3   Violation in state 4

Fig. 2. Step 4.2

Overall Correctness. Step 1 only changes the labels of states of a PKS. Thus, the probability space of the PKS is unchanged. Hence, after projecting $\mathcal{A}_\delta$ on l, we can reformulate SSPOD-1 in terms of $\mathcal{A}_{\delta\,|_l}$. Let $(\Omega, \mathcal{F}, \mathbf{P}_{\delta,l})$ denote the probability space of $\mathcal{A}_{\delta\,|_l}$. First, we reformulate SSPOD-1, which talks about projected traces in $\mathcal{A}$, in terms of the traces in the projected $\mathcal{A}_{\delta,l}$

Theorem 2. *For any $l \in L$, and for a set of traces $X \in \mathcal{F}$ that are closed under stuttering equivalence, if $\mathbf{P}_{\delta,l}[X] = 1$ or $\mathbf{P}_{\delta,l}[X] = 0$, then SSPOD-1 holds.*

Proof. See Appendix C of [20].

The key step (Step 2) in our algorithm is the reduction of a probabilistic property to a non-probabilistic property: after removing all stuttering loops, if all traces of $\mathcal{A}_{\delta\,|_l}$ are stuttering equivalent, then $\mathbf{P}_{\delta,l}[X] = 1$. Thus, SSPOD-1 holds. The correctness of this step follows from a result from Baier and Kwiatkowska [5]: whenever *all fair traces* of a PKS fulfill a certain property φ, then φ holds *with probability* 1. In our context, we define the fairness of traces w.r.t. *non-stuttering transitions*. A non-stuttering transition is *enabled* in a state T_i iff there exists a finite sequence of transitions from T_i that leads to T_j such that $V(T_j) \neq V(T_i)$. A non-stuttering transition is said to be *taken* in a state T_i of T iff $\exists j > 0.\, T_i \neq T_{i+j}$. A trace is *strongly fair* w.r.t. non-stuttering transitions if given that a non-stuttering transition is enabled infinitely often, it is taken infinitely often. Thus, a trace that stutters in a non-final stuttering loop forever is unfair. Let $Fair(\mathcal{A})$ denote the set of fair traces of $Trace(\mathcal{A})$. Applying the result from [5], we obtain:

Theorem 3. *Given a finite $\mathcal{A}_{\delta\,|_l}$ and a set of traces $X \in \mathcal{F}$ that are closed under stuttering equivalence and do not stutter forever in a non-final stuttering loop, if $\forall T, T' \in Fair(\mathcal{A}_{\delta\,|_l}).\, T \sim T'$, then $\mathbf{P}_{\delta,l}[X] = 1$.*

We show that after removing all stuttering loops, and re-establishing the self-loops for final states, the set of fair traces of $\mathcal{A}$ is preserved.

Theorem 4. *Given a PKS $\mathcal{A}$, let $\mathcal{R}$ denote the PKS that is obtained after removing all stuttering loops and re-establishing the self-loops for final states. Then $Fair(\mathcal{A}) = Trace(\mathcal{R})$.*

Proof. See Appendix D of [20].

Combining these results, we obtain.

Theorem 5. *For any $l \in L$, if all traces of $\mathcal{R}_{\delta\,|_l}$ are stuttering equivalent, then SSPOD-1 holds.*

Overall Complexity. Step 1 labels every state of $\mathcal{A}$ by the value of l in that state. This is done in time complexity $O(n)$, where n is the number of states of $\mathcal{A}$. Step 2 uses an $O(m)$-algorithm to find the strongly connected components, where m is the number of transitions of $\mathcal{A}$. The time complexity of Step 4.1 is also $O(m)$. The core of Step 4.2 is the BFS algorithm, whose running time is $O(n+m)$. Therefore, for a single low variable l, the total time complexity of the verification is linear in the size of $\mathcal{A}$, i.e., $O(n+m)$, and for any initial state, the total complexity of the verification of SSPOD-1 (for all $l \in L$) is $|L|\,O(n+m)$.

4.2   Verification of SSPOD-2

Algorithm. SSPOD-2 states that, given a program C, for any two initial low-equivalent states I and I', if we project on the set of low variables L, the probabilistic languages arising from the executions of I and I' should be the same. A number of efficient algorithms for checking equivalence between probabilistic languages have been developed, the classical ones in [10,29], and the improved variants in [11,16]. However, none of the existing algorithms exactly fit our purposes, since either they do not abstract from stuttering steps [29,11,16], or they consider a different variation of probabilistic language inclusion [10].

Therefore, to verify SSPOD-2, our algorithm first transforms the PKS into an equivalent one, without *stuttering steps*, and then we use the latest and most efficient algorithm from Kiefer et al. [16] to check equivalence of these probabilistic languages. The basic idea of this algorithm is to present the language of a PKS by a polynomial in which each monomial presents an input word of the language and the coefficient of the monomial represents the weight of the word, i.e., the probability of the execution of the word. This method reduces the language equivalence problem to polynomial identity testing.

————SSPOD-2————————————————————————————

1: Project both $\mathcal{A}_\delta$ and $\mathcal{A}'_\delta$ (modeling the executions starting in I and I') on the set L, yielding $\mathcal{A}_{\delta\,|_L}$ and $\mathcal{A}'_{\delta\,|_L}$.

2: Remove all stuttering steps from $\mathcal{A}_{\delta\,|_L}$ and $\mathcal{A}'_{\delta\,|_L}$, yielding stuttering-free PKSs $\mathcal{R}_{\delta\,|_L}$ and $\mathcal{R}'_{\delta\,|_L}$.

3: Check the equivalence of the stuttering-free probabilistic languages between $\mathcal{R}_{\delta\,|_L}$ and $\mathcal{R}'_{\delta\,|_L}$, using Kiefer et al. [16].

Overall Correctness. After projecting both $\mathcal{A}_\delta$ and $\mathcal{A}'_\delta$ on L, we can reformulate SSPOD-2 in terms of $\mathcal{A}_{\delta|_L}$ and $\mathcal{A}'_{\delta|_L}$. Let $(\Omega, \mathcal{F}, \mathbf{P}_{\delta,L})$ and $(\Omega, \mathcal{F}, \mathbf{P}'_{\delta,L})$ denote the probability space of $\mathcal{A}_{\delta|_L}$ and $\mathcal{A}'_{\delta|_L}$, respectively.

Theorem 6. *SSPOD-2 holds iff for all sets of traces $X \in \mathcal{F}$ that are closed under stuttering equivalence, we have $\mathbf{P}_{\delta,L}[X] = \mathbf{P}'_{\delta,L}[X]$.*

Proof. See Appendix E of [20].

Let $\mathcal{R}$ denote a stuttering-free PKS that is obtained by applying Step 2 on a given $\mathcal{A}$. Let $P_\mathcal{A}$ and $P_\mathcal{R}$ be the probabilistic transition functions of $\mathcal{A}$ and $\mathcal{R}$, respectively. Step 2 removes all stuttering steps by changing $P_\mathcal{A}$ to $P_\mathcal{R}$ given by the following equations.

$$P_\mathcal{R}(c,c') = \begin{cases} P_\mathcal{A}(c,c') & \text{if } V(c) \neq V(c') \\ \sum_{c'':V(c)=V(c'')} P_\mathcal{A}(c,c'') \, P_\mathcal{R}(c'',c') & \text{otherwise.} \end{cases}$$

Thus, for non-stuttering steps, $P_\mathcal{A}$ and $P_\mathcal{R}$ are the same; for stuttering steps, $P_\mathcal{R}$ accumulates the probabilities of moving to c' via some stuttering steps $c \to c''$. Thus, $P_\mathcal{R}$ accounts for the transition probabilities of stuttering steps in $\mathcal{A}$ into the transition probabilities of non-stuttering steps in $\mathcal{R}$. Therefore, removing stuttering steps does not change the probabilities of sets of traces that are closed under stuttering equivalence.

Theorem 7. *Let $X \in \mathcal{F}$ be a set of traces that are closed under stuttering equivalence, then $\mathbf{P}_\mathcal{A}[X] = \mathbf{P}_\mathcal{R}[X]$.*

Combining all results, it is obvious that to check SSPOD-2, we can check for probabilistic language equivalence between $\mathcal{R}_{\delta|_L}$ and $\mathcal{R}'_{\delta|_L}$.

Overall Complexity. Step 1 is done in time complexity $O(n)$, where n is the number of states of two PKSs. Step 2 essentially calculates a reachability probability, and is defined as a system of n linear equations over n variables. This equation system can be solved in $O(n^3)$. Step 3 can be done in $O(nm)$, where m is the number of transitions [16]. Thus, the overall complexity is $O(n^3)$ for each pair I and I'.

5   Related Work

The idea of observational determinism originates from Roscoe [23], who was the first to identify the need for determinism to ensure confidentiality for concurrent processes. This observation has resulted in several subtly different definitions of observational determinism for possibilistic multi-threaded programs (e.g., [31,15,28,14]), see [14] for a detailed comparison. Notice that, SSOD is the only one to consider the effect of schedulers on confidentiality.

When programs have probabilistic behaviors, to prevent information leakage under probabilistic attacks, several notions of probabilistic noninterference have been proposed [30,24,25]. The first is from Volpano and Smith [30]. It is based on a lock-step execution of probability distributions on states, i.e., given any two initial states that are indistinguishable w.r.t. low variables, the executions of the program from these two initial states, after projecting out high variables, are exactly the same. As shown by Sabelfeld and Sands [24], this definition is not precise, and overly restrictive. Sabelfeld and Sands's definition of probabilistic noninterference is based on a *partial probabilistic low-bisimulation* [24], which requires that given any two initial states that are indistinguishable w.r.t. low variables, for any trace that starts in an initial state, *there exists* a trace that starts in the other initial state and passes through the same equivalence classes of states at the same time, with the same probability. This definition is restrictive *w.r.t.* timing, i.e., it cannot accommodate threads whose running time depends on high variables. Thus, it rejects many harmless programs, while SSPOD accepts, such as if (h > 0) then {l1 := 3; l1 := 3; l2 := 4} else {l1 := 3; l2 := 4}.

To overcome these limitations, Smith proposes to use a weak probabilistic bisimulation [25]. Weak probabilistic bisimulation allows two traces to be equivalent when they reach the same outcome, but one runs slower than the other. However, this still demands that any two bisimilar states must reach indistinguishable states with the same probability. This condition of probabilistic bisimulation is more restrictive than SSPOD, because when trace occurrences do not depend on high variables, probabilistic noninterference still rejects the program.

Moreover, all bisimulation-based definitions mentioned above do not require the deterministic behavior of each low variable. However, we insist that a multithreaded program must enforce a deterministic orderings on the accesses to low variables, see [14]. Finally, probabilistic noninterference [24,25] also put restrictions on unreachable states, e.g., l := 1; if (l == 0) then l := h else skip is secure but rejected, because the bisimulation also considers the case when the conditional statement is executed from an unreachable state where l equals 0, see [8]. Mantel et al. [18] overcome this limitation by explicitly using assumptions and guarantees about how threads access the shared memory. Notice that SSPOD does not have this property, thus SSPOD is less restrictive.

Mantel et al. [19] also consider the effect of schedulers on confidentiality. However, their observational model is different from ours. They assume that the attacker can only observe the initial and final values of low variables on traces. Thus, their definitions of confidentiality are noninterference-like.

Palamidessi et al., Chen et al., Smith, and Zhu et al. [2,3,4,9,26,32] investigate a quantitative notion of information leakage for probabilistic systems. Quantitative analysis offers a method to compute *bounds* on how much information is leaked. This information can be used to compare with the threshold, and thus suggesting whether the program is accepted or not. Therefore, we can tolerate the *minor* leakage. Thus, this line of researches is complementary to ours.

6   Conclusion

Summary. This paper introduces the notion of *scheduler-specific probabilistic observational determinism*, together with an algorithmic verification technique.

SSPOD captures the notion of confidentiality for probabilistic multithreaded programs. The definition extends an earlier proposal for possibilistic confidentiality of such programs, and makes it usable in a larger context. It is important to consider probabilistic multi-threaded programs, because this captures the realistic behavior of programs.

We also propose an algorithmic verification technique for it. The verification is using a combination of new and existing algorithms. The new algorithm solves a standard problem, which makes it applicable also in a broader context. We believe that the idea of adapting known model checking algorithms will also be appropriate for other security properties, such as integrity and availability.

Future Work. We see several directions for future work. We plan to continue the study of other security properties, i.e., anonymity, integrity, and availability. We believe that our algorithmic approach is also appropriate to efficiently and precisely verify these security properties.

Further, we also plan to relax our definitions of confidentiality by quantifying the information flow and determining how much information is being leaked. The existing models of quantitative analysis do not address which measure is suitable to quantify information leakage for multi-threaded programs, thus a new approach has to be developed.

Acknowledgment. Our work is supported by NWO under grant 612.067.802 (SLALOM) and grant Dn 63-257 (ROCKS).

References

1. Aho, A.V., Hopcroft, J.E.: The Design and Analysis of Computer Algorithms, 1st edn. Addison-Wesley (1974)
2. Alvim, M.S., Andrés, M.E., Chatzikokolakis, K., Palamidessi, C.: On the Relation between Differential Privacy and Quantitative Information Flow. In: Aceto, L., Henzinger, M., Sgall, J. (eds.) ICALP 2011, Part II. LNCS, vol. 6756, pp. 60–76. Springer, Heidelberg (2011)
3. Alvim, M.S., Andrés, M.E., Chatzikokolakis, K., Palamidessi, C.: Quantitative Information Flow and Applications to Differential Privacy. In: Aldini, A., Gorrieri, R. (eds.) FOSAD VI 2011. LNCS, vol. 6858, pp. 211–230. Springer, Heidelberg (2011)
4. Andres, M.E., Palamidessi, C., Sokolova, A., Van Rossum, P.: Information hiding in probabilistic concurrent systems. Journal of Theoretical Computer Science 412(28), 3072–3089 (2011)
5. Baier, C., Kwiatkowska, M.: On the verification of qualitative properties of probabilistic processes under fairness constraints. Information Processing Letters 66, 71–79 (1998)
6. Barthe, G., D'Argenio, P., Rezk, T.: Secure information flow by self-composition. In: CSFW, pp. 100–114. IEEE Press (2004)
7. Blom, S., van de Pol, J., Weber, M.: LTSMIN: Distributed and Symbolic Reachability. In: Touili, T., Cook, B., Jackson, P. (eds.) CAV 2010. LNCS, vol. 6174, pp. 354–359. Springer, Heidelberg (2010)
8. Blondeel, H.-C.: Security by logic: characterizing non-interference in temporal logic. Master's thesis, KTH Sweden (2007)
9. Chen, H., Malacaria, P.: Quantitative analysis of leakage for multi-threaded programs. In: PLAS 2007 (2007)

10. Christoff, L., Christoff, I.: Efficient Algorithms for Verification of Equivalences for Probabilistic Processes. In: Larsen, K.G., Skou, A. (eds.) CAV 1991. LNCS, vol. 575, pp. 310–321. Springer, Heidelberg (1992)
11. Doyen, L., Henzinger, T.A., Raskin, J.F.: Equivalence of labeled Markov chains. Int. J. Found. Comput. Sci. 19(3), 549–563 (2008)
12. Goguen, J.A., Meseguer, J.: Security policies and security models. In: IEEE Symposium on Security and Privacy (1982)
13. Gurfinkel, A., Chechik, M.: Why Waste a Perfectly Good Abstraction? In: Hermanns, H., Palsberg, J. (eds.) TACAS 2006. LNCS, vol. 3920, pp. 212–226. Springer, Heidelberg (2006)
14. Huisman, M., Ngo, T.M.: Scheduler-Specific Confidentiality for Multi-threaded Programs and Its Logic-Based Verification. In: Beckert, B., Damiani, F., Gurov, D. (eds.) FoVeOOS 2011. LNCS, vol. 7421, pp. 178–195. Springer, Heidelberg (2012)
15. Huisman, M., Worah, P., Sunesen, K.: A temporal logic characterization of observation determinism. In: CSFW. IEEE Computer Society (2006)
16. Kiefer, S., Murawski, A.S., Ouaknine, J., Wachter, B., Worrell, J.: Language Equivalence for Probabilistic Automata. In: Gopalakrishnan, G., Qadeer, S. (eds.) CAV 2011. LNCS, vol. 6806, pp. 526–540. Springer, Heidelberg (2011)
17. Kripke, S.A.: Semantical considerations on modal logic. Acta Philosophica Fennica 16, 83–94 (1963)
18. Mantel, H., Sands, D., Sudbrock, H.: Assumptions and guarantees for compositional noninterference. In: CSF 2011, pp. 218–232 (2011)
19. Mantel, H., Sudbrock, H.: Flexible Scheduler-Independent Security. In: Gritzalis, D., Preneel, B., Theoharidou, M. (eds.) ESORICS 2010. LNCS, vol. 6345, pp. 116–133. Springer, Heidelberg (2010)
20. Ngo, T.M., Stoelinga, M., Huisman, M.: Confidentiality for probabilistic multithreaded programs and its verification. Full version, http://wwwhome.ewi.utwente.nl/~ngominhtri/
21. Ngo, T.M., Stoelinga, M., Huisman, M.: Effective verification of confidentiality for multi-threaded programs. Manuscript 201X
22. Peled, D., Wilke, T.: Stutter-invariant temporal properties are expressible without the next-time operator. Information Processing Letters 63, 243–246 (1997)
23. Roscoe, A.W.: CSP and determinism in security modeling. In: IEEE Symposium on Security and Privacy, pp. 114–127. IEEE Computer Society (1995)
24. Sabelfeld, A., Sands, D.: Probabilistic noninterference for multi-threaded programs. In: CSFW, pp. 200–214 (2000)
25. Smith, G.: Probabilistic noninterference through weak probabilistic bisimulation. In: CSFW (2003)
26. Smith, G.: On the Foundations of Quantitative Information Flow. In: de Alfaro, L. (ed.) FOSSACS 2009. LNCS, vol. 5504, pp. 288–302. Springer, Heidelberg (2009)
27. Stoelinga, M.I.A.: Alea jacta est: verification of probabilistic, real-time and parametric systems. PhD thesis, University of Nijmegen, The Netherlands (April 2002)
28. Terauchi, T.: A type system for observational determinism. In: CSF (2008)
29. Tzeng, W.G.: A polynomial-time algorithm for the equivalence of probabilistic automata. SIAM Journal on Computing 21, 216–227 (1992)
30. Volpano, D., Smith, G.: Probabilistic noninterference in a concurrent language. Journal of Computer Security 7, 231–253 (1999)
31. Zdancewic, S., Myers, A.C.: Observational determinism for concurrent program security. In: CSFW, pp. 29–43. IEEE (2003)
32. Zhu, J., Srivatsa, M.: Quantifying information leakage in finite order deterministic programs. In: CoRR 2010 (2010)

A Fully Homomorphic Crypto-Processor Design
Correctness of a Secret Computer

Peter T. Breuer[1,*] and Jonathan P. Bowen[2,**]

[1] Department of Computer Science, University of Birmingham, UK
ptb@cs.bham.ac.uk
[2] Faculty of Business, London South Bank University, UK
jonathan.bowen@lsbu.ac.uk

Abstract. A KPU is a replacement for a standard CPU that natively runs encrypted machine code on encrypted data in registers and memory – a 'crypto-processor unit', in other words. Its computations are opaque to an observer with physical access to the processor but remain meaningful to the owner of the computation. In theory, a KPU can be run in simulation and remain as secure (or otherwise) as in hardware. Any block cipher with a block-size of about a word is compatible with this developing technology, the long-term aim of which is to make it safe to entrust data-oriented computation to a remote environment.

Hardware is arranged in a KPU to make the chosen cipher behave as a mathematical homomorphism with respect to computer arithmetic. We describe the architecture formally here and show that 'type-safe' programs run correctly when encrypted.

1 Introduction

A KPU is a replacement for a standard CPU ('central processor unit') that natively runs encrypted machine code on encrypted data in registers and memory. The term 'KPU' is derived from 'crypto-processor', and while the latter has been used for several hardware-based units aimed at helping overall system security (see, for example, [3, 10, 15]), we mean it in the literal sense of a complete general purpose processor that has been architected to perform all its computations encrypted. Any block cipher is compatible provided that the block-size is not impractical – it dictates the physical size of an information word. The technology is aimed at allowing data-oriented applications such as fluid dynamics computations, image processing, even cryptography, to run in an insecure environment in relative security.

An observer can recognize control flow (jumps, branches, etc.), but the meaning of the data is hidden by the encryption. An observer may see the calculation

$$43 \ \# \ 43 = 21234089$$

but that it represents $1 + 1 = 2$ is known only to the owner of the computation.

* Some work reported here was carried out as Visiting Fellow at LSBU.
** Jonathan Bowen acknowledges the support of Museophile Limited.

J. Jürjens, B. Livshits, and R. Scandariato (Eds.): ESSoS 2013, LNCS 7781, pp. 123–138, 2013.

Everything a programmer needs to know about a KPU is summarized in Box 1. The machine code is an encrypted form of the standard RISC [11] instruction set. The interested reader will find an 'on the metal' instruction format in the patent [1], which sets out design rules that result in a correctly working KPU, whatever the block cipher chosen, plus a reference implementation.

> **Box 1. A KPU** is a CPU that natively processes mixed encrypted and unencrypted data in general purpose registers and memory. It executes encrypted RISC machine code instructions on:
>
> - encrypted data and data addresses,
> - unencrypted program addresses.
>
> Instructions must be programmed with encryption type in mind. Arithmetic instructions apply to encrypted data, jumps to unencrypted program addresses. But ...
>
> - memory load/store instructions are *polymorphic*: they copy data whether encrypted or unencrypted.

This paper deals with correctness. As well as setting out the KPU design and design principles, we show that a KPU running an encrypted machine code program runs 'correctly': it generates machine states that are encryptions of the machine states expected in an ordinary RISC CPU running the corresponding unencrypted machine code program. The only proviso is that the running program does not 'break the conventional apartheid between program and data', which we characterize as *type-safe for a KPU* (also known as 'crypto-safe') below. That means in particular that a compiler for the KPU needs to be run outside of the KPU.

Definition 1. *A program is* type-safe *for a KPU ('crypto-safe') if those KPU machine instructions that work on encrypted data always get encrypted data on which to work during execution of the program, while those instructions that work on unencrypted data always get unencrypted data on which they can work.*

The need for such a notion arises from the fact that a *mix* of encrypted and unencrypted data is always circulating inside a KPU and through memory and registers. While data and data addresses are encrypted, program addresses are not. More generally, program address encryption needs to be different from data and data address encryption, but we will suppose the program address encryption is null for the purposes of this exposition. On the one hand the distinction is physically mandated: the circuit that updates the program counter is distinct from the circuit that does the general arithmetic, so the encryptions may be different without interfering. On the other hand the two encryptions ought to be different for cryptographic reasons: the usual change in the program counter from cycle to cycle is an increment by 4 on a 32-bit machine, so valuable information could be garnered were the counter to be observed.

Type-safety in the KPU is explored in more detail in [2]. The special KPU RISC+CRYPT assembly is a typed language, and those assembly language programs that type-check correctly are shown in [2] to be type-safe for the KPU.

This paper is structured as follows: first, a top-down view of KPU design is given in Section 2 and 3, then we give a description of a RISC CPU in Section 4,

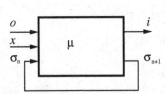
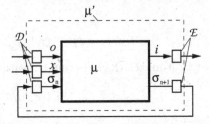

Fig. 1. An abstract view of a processor with transition function μ, sending inputs i to peripherals and receiving responses o when treating instruction x.

Fig. 2. Adding imaginary codecs $\mathcal{D}$, $\mathcal{E}$ that cancel with $\mathcal{D} \circ \mathcal{E} = I$ to an abstract processor. μ' marks out a processor that works with encrypted state and I/O.

detailing a KPU further in Section 5. We compare the two in Section 6, ultimately deriving the asserted 'correct' correspondence between the behaviours.

2 Overview

This section gives a high-level view of KPU design principles. A current-day CPU can be regarded in the abstract as a finite state machine μ with inputs o and outputs i that can be visualized as a black-box as in Fig. 1. The state σ managed by the state machine consists of the registers of the CPU. That may consist of a 32-bit value at each of approximately 32 registers within the processor, while the inputs and outputs constitute the processor's communications with peripherals (counting 'random access memory' – RAM – as a peripheral). We view the processor outputs i as 'inputs' (commands) to the peripherals and the processor inputs o as 'outputs' (responses) from the peripherals.

Formally, on the nth clock cycle, the processor is executing the instruction x, say, and effects the following transformation:

$$\mu_x(\sigma_n, o) = (\sigma_{n+1}, i)$$

where μ is the 'state transition function', here written subscripted by the instruction x being executed. Where convenient we shall write $\mu(-, x, -)$ for μ_x.

One may introduce encryption into this picture by imagining codecs on both sides of the processor, as shown in Fig. 2. Encryption is denoted by units labelled $\mathcal{E}$ and decryption by units labelled $\mathcal{D}$, with the composition $\mathcal{D} \circ \mathcal{E}$ being the identity. Nothing changes from the point of view of the processor μ. It sees the same states as always since the pair of codecs in the state feedback loop cancel out – they might as well not be there at all as far as the processor is concerned. If it sees the same inputs o as always, it produces the same outputs i. It is only up to us to provide appropriately encrypted inputs at left in order that the processor continue to see the same o and x on the right of the decoder $\mathcal{D}$.

The mathematical content of the above paragraph is as follows. Letting σ' be the encrypted state, with $\mathcal{D}(\sigma') = \sigma$, and x' be the encrypted instruction, with $\mathcal{D}(x') = x$, we see that the dotted lines in Fig. 2 delimit a processor with transition function μ', related to μ by

$$(\mathcal{D} \times \mathcal{D}) \circ \mu'_{x'} = \mu_x \circ (\mathcal{D} \times \mathcal{D}) \tag{1}$$

That is, it makes the following transition on the nth cycle:

$$\mu'_{x'}(\sigma'_n, o') = (\sigma'_{n+1}, i')$$

where i' and o' are respectively encrypted inputs to/outputs from peripherals.

When we provide the appropriate sequence of inputs o and o' respectively, the two systems evolve as one. So if one builds a machine with transition function μ' instead of μ, it will behave perfectly correctly, but inputs and outputs and states will be encrypted. How can one build it?

It turns out that if we lay out any standard RISC processor hardware design without integrating too early the computer arithmetic part, then it can be done. Though we will not work through the design transformation, it means pushing the imaginary codecs in the diagram of Fig. 2 deeper and deeper into the hardware until they meet up and cancel. As imaginary codecs are 'pushed through' each combinatorial logic unit inside the CPU, they leave behind on the drawing board a unit with altered functionality bearing the same relationship to the original as the altered transition function μ' bears to its original μ via (1).

We will describe this transformed design mathematically, and prove it runs correctly. The presentation is necessarily abstract, but all the details left out can be filled in from the instruction manual [8] for a RISC processor.

3   KPU versus RISC CPU

In this section, we aim to sketch out the relation that holds between a KPU (crypto-processor) and a reference RISC CPU when running the same program. By the end of the paper we will have proved the result that is only stated here.

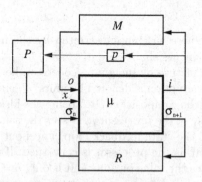

Fig. 3. Adding heap memory M, program counter p, program memory P and registers R to the abstract picture of a CPU

A RISC CPU has just *five* parts: R, p, M, P, μ. The internal state is represented by the component R, a vector consisting of the contents of the 32 32-bit general registers. The components M, P, p are external to the processor chip. The first, M, represents the read-write memory area, and the second, P, represents the read-only program memory area. The third, p, is considered peripheral to the CPU here for the purposes of this exposition, whereas in terms of physical design and proximity it is most certainly an internal component. It is the program counter register content. The processor needs to refer to this value every cycle in order to fetch the right instruction from memory for decoding. A

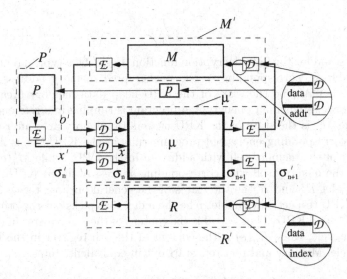

Fig. 4. Designing a KPU by 'pushing through' cancelling pairs of imaginary codecs into a CPU abstract schematic. Dotted lines demarcate KPU components

finite state machine transition function μ representing combinatorial logic in the processor completes the list. Fig. 3 shows their relationship, refining Fig. 1.

A KPU, abstractly, has the corresponding altered parts R', p', M', P' and μ' respectively, as shown in Fig. 4, where the derivation by 'pushing through' imaginary codec pairs in Fig. 3 is illustrated. KPU data memory M' behaves just like CPU memory M containing encrypted data at encrypted addresses (byte-wise access is implemented via arithmetic on full words) because the imaginary codec on input to M decodes both address and data lines. The KPU's program memory P' behaves like the CPU memory P with encrypted instructions stored at unencrypted addresses. The KPU registers R' behave like CPU registers R with encrypted data stored at unencrypted locations – the imaginary codec on input to R decodes data but not index lines. Thus ordinary register and memory components are used in a KPU.

We set both KPU and CPU up initially so that encrypted program instructions lie in memory in a KPU at precisely those addresses where the corresponding unencrypted program instructions lie in a CPU. That is, $P(p)$ is the instruction at program address p in the CPU and $P'(p)$ is the encryption of that instruction at the same program address p in the KPU. Let $\mathcal{D}$ stand for the instruction decryption function, then

$$P(p) = \mathcal{D}(P'(p)) \qquad (2)$$

holds initially. Moreover, we consider the program areas P and P' to be read-only, so (2) stays that way through program execution.

Let the value in the program counter register **pc** be p_n and p'_n respectively in CPU and KPU after n cycles. The program counter starts off pointing to the same entry point in both CPU and KPU:

$$p_0 = p_0' \tag{3}$$

Now let $\mathcal{D}$ stand for the data decryption function. Let a' be a typical (encrypted) memory address, corresponding to the unencrypted memory address a, with $\mathcal{D}(a') = a$. Let M_n be the state of the CPU heap data memory area after n instruction cycles when an unencrypted program is running in a reference RISC CPU, while M_n' is the state of the KPU heap data memory area after n cycles when the corresponding encrypted program runs in the KPU. Thus $M_n(a)$ is the content of the memory cell with address a in the CPU, while $M_n'(a')$ is the content of the memory cell with corresponding address a' in the KPU.

Similarly let R_n and R_n' be the states of the general purpose registers in the CPU and KPU respectively, and let ρ be the index that picks out the ρth register in both CPU and KPU. Thus $R_n(\rho)$ is the content of the ρth register in the CPU on the nth cycle, while $R_n'(\rho)$ is the content of the ρth register in the KPU on the nth cycle. Memory and registers start out in equivalent states:

$$M_0(a) = \mathcal{D}(M_0'(a')) \tag{4}$$
$$R_0(\rho) = \mathcal{D}(R_0'(\rho)) \tag{5}$$

We are now able to state the result that this paper aims at.

Theorem 1. *For a type-safe (aka 'crypto-safe') program running in both CPU and KPU, provided the program loader sets up equivalent initial states as specified by (2–5), the states obtained thereafter are the same modulo encryption:*

$$M_n(a) = \mathcal{D}(M_n'(a')) \tag{6}$$
$$R_n(\rho) = \mathcal{D}(R_n'(\rho)) \tag{7}$$
$$p_n = p_n' \tag{8}$$

In the following pages both CPU and KPU are described as satisfying deterministic systems of recursion equations that are transformed to one another via a relation extending (6–8), which proves Theorem 1.

In practice, the initial conditions (4–5) need hold only for those addresses and registers that are read before they are written during execution. Since it should never be the case that an application program reads dynamic memory before writing it, we can forget about (4) in practice, except for the encrypted addresses a' of the (encrypted) read-only constants that the program loader embeds in memory prior to program execution. Likewise, registers should not be read before written, so one may forget about (5) with the exception of register number $\rho = 0$, which is fixed at zero in a RISC machine, and which may well be read first by programs which rely on that.

To illustrate the predictive power of Theorem 1 we prove an immediate corollary. The special instruction register **ir** in a CPU always contains a copy

$$x_n = P(p_n) \tag{9}$$

of the instruction x_n found at address p_n (the program counter contents at cycle n) in the CPU program memory area P, and in the KPU it contains a copy

$$x'_n = P'(p'_n) \tag{9'}$$

of the instruction x'_n at address p'_n in the KPU program memory area P'.

Corollary 1. *The* **ir** *contents x_n in a CPU and x'_n in a KPU at cycle n always correspond during runs of a type-safe machine code program (and its encryption, respectively) in CPU and KPU, modulo the encryption. That is:*

$$x_n = \mathcal{D}(x'_n)$$

Proof. That is by virtue of (2) and (8) with (9) and (9'). □

Note that the decryption function $\mathcal{D}$ implicitly varies from place to place. We should write $\mathcal{D}_P(P'(p'))$ and $\mathcal{D}_M(M'_n(a'))$, but we take the subscripts as read.

4   Formal Description of a RISC CPU

This section provides the abstract view of a reference RISC CPU that serves as a basis for comparison with a KPU (crypto-processor). It is a succinct description, and therefore necessarily dense, but we hope that familiarity with the real-life model leaves the reader able to concentrate on the special features.

The first special feature is that we view memory and program counter as separate to the CPU proper, working in parallel as an independent, joint communicating process that receives requests i and returns responses o. Requests i emitted by the processor are of type $\mathbf{I}$, responses o received are of type $\mathbf{O}$, and these I/O commands refer to memory elements sited at addresses of type $\mathbf{D}$.

The type $\mathbf{D}$ is a general type for data words (32-bit binary values, in practice), so we are saying that addresses are just one kind of data and the processor does not inherently know the difference, which is true. The processor may also, in principle, confuse data addresses and program addresses because they look the same, nevertheless we choose to pick out and name the subset $\mathbf{P}$ of instruction addresses $\mathbf{P} \subseteq \mathbf{D}$ in order to make certain type expressions look more meaningful to the reader. We also set $\mathbf{X}$ to be the type of program instructions. Again, in practice, this is just another special subset $\mathbf{X} \subseteq \mathbf{D}$ of the type $\mathbf{D}$ of 32-bit data words, but we pick it out and name it in order to enhance readability here.

Set $\mathbf{R} = \mathbf{D}^{32}$, then the registers state R in a RISC CPU comes from domain $\mathbf{R}$, the program counter value p comes from domain $\mathbf{P}$, memory outputs o from domain $\mathbf{O}$, memory inputs i from domain $\mathbf{I}$, program instructions x from domain $\mathbf{X}$. On the nth cycle, the transition effected by the CPU is as follows:

$$\mu : \mathbf{R} \times \mathbf{X} \times \mathbf{O} \to \mathbf{R} \times \mathbf{I}$$
$$(R_{n+1}, i_n) = \mu(R_n, x_n, o_n) \tag{10}$$

The reason why the request i_n to memory and the response o_n on different sides of the equation get the same subscript is that memory (via cache) reacts within one cycle, and the suffix merely denotes that cycle. This view is apparently simplistic – in reality memory accesses sometimes take longer than a single cycle to complete even though programs are carefully arranged by the compiler so that never has an effect – but it simplifies the presentation. Without it, we would have to split the transition function into two halves, one to emit the request and one to treat the response, and the technicality would obscure the main issues.

Requests $i : \mathbf{I}$ emitted by the CPU and responses $o : \mathbf{O}$ received by the CPU have the forms below:

$$\mathbf{I} = \mathbf{D} \times \mathbf{D} \sqcup \mathbf{D} \sqcup \mathbf{P} \sqcup \mathbb{1} \qquad \mathbf{O} = \mathbf{D} \sqcup \mathbb{1}$$
$$i = \ a!d \ \mid a? \mid \uparrow p \mid \phi \qquad o = d \mid \phi$$

Requests of the form $a!d$ and $a?$ are, respectively, write and read requests to memory for accesses at address a. The request $\uparrow p$ is directed to the program counter, commanding it change to program address p. Request ϕ is 'no request'. The empty response ϕ is received in every case except a read request $a?$ to memory, when the datum d at address a is received in response.

We need to formalise the memory and program counter behaviour in terms of requests and responses. Let $M_n : \mathbf{D} \to \mathbf{D}$ be the data memory at cycle n, and $P : \mathbf{P} \to \mathbf{X}$ be the program memory – which we suppose read-only, hence constant during a program run. The following equations define memory and program counter behaviour according to ordinary intuition. Memory changes according to a write request $i_n = a!d$ as in (11) and responds to a read request $i_n = a?$ as in (12). The program counter changes in response to a jump request $i_n = \uparrow p$ as in (13), otherwise it is incremented each cycle by the size of an instruction word, which we abbreviate as '4' to avoid too much confusing generality:

$$M_{n+1}(a) = \begin{cases} d & i_n = a!d \\ M_n(a) & \textbf{otherwise} \end{cases} \tag{11}$$

$$o_n = \begin{cases} M_n(a) & i_n = a? \\ \phi & \textbf{otherwise} \end{cases} \tag{12}$$

$$p_{n+1} = \begin{cases} p & i_n = \uparrow p \\ p_n + 4 & \textbf{otherwise} \end{cases} \tag{13}$$

The behaviour of memory and program counter in (11–13) above in conjunction with (10) and also the statement (9) that the nth program instruction is fetched from program memory at the address given by the program counter on the nth cycle completely specifies RISC CPU behaviour. It remains only to refine the processor transition function μ of (10) to detail the behaviour per instruction.

The specification of μ may be extracted from the manual [8]. The cases are each predicated on the functionality Λ of an Arithmetic Logic Unit (ALU) and $\varXi$ of a Sign Extension Unit (SEU) within the RISC CPU (the SEU is responsible for the embedding of 16-bit numbers in 32-bit space). For each function code ι, Λ_ι produces a 'result and carry' pair from two operands:

$$\Lambda_\iota : D \times D \to D \times D \qquad\qquad \Xi : D \to D$$

We will not go through each RISC instruction here. Here are three exemplars:

1. The instruction $x_n = \mathbf{addiu}\,\rho_1\,\rho_2\,k$ invokes the following transition, where π_1 is the projection to the first of the pair of outputs from the ALU:

$$R_{n+1}(\rho) = \begin{cases} \pi_1(\Lambda_{\mathrm{add}}(R_n(\rho_2), \Xi(k))) & 0 \neq \rho = \rho_1 \\ R_n(\rho) & \textbf{otherwise} \end{cases} \qquad (14)$$

$$o_n = i_n = \phi$$

The '$0 \neq \rho_1$' expresses that, in a RISC CPU, register zero contains an immutable (zero) value, so writing to it does nothing. Otherwise, the instruction adds k to the content of register ρ_2 and writes the result in register ρ_1.

2. A jump and link instruction $x_n = \mathbf{jal}\,p$ invokes the following transition:

$$R_{n+1}(\rho) = \begin{cases} p_n + 4 & \rho = \mathbf{ra} \\ R_n(\rho) & \textbf{otherwise} \end{cases} \qquad (15)$$

$$i_n = \uparrow p, \qquad o_n = \phi$$

We have simplified here – in reality only 26 bits of the jump target address p are supplied in the instruction, the top bits being filled in from the instruction address p_n. The address $p_n + 4$ of the next instruction is written to the $\mathbf{ra}$ ('return address') register.

3. A load word instruction $x_n = \mathbf{lw}\,\rho_1\,k(\rho_2)$ invokes the following transition:

$$R_{n+1}(\rho) = \begin{cases} d & 0 \neq \rho = \rho_1 \\ R_n(\rho) & \textbf{otherwise} \end{cases} \qquad (16)$$

$$i_n = a? \quad \textbf{where}\ a = \pi_1(\Lambda_{\mathrm{add}}(R_n(\rho_2), \Xi(k)))$$

$$o_n = d \qquad\qquad d = M_n(a)$$

This instruction fetches from memory at a location a offset by k from the address in register ρ_2, and writes to register ρ_1.

One may have confidence the full specification is correct as it runs RISC machine code correctly.

5   Formal Description of a KPU

This section describes a KPU (crypto-processor). There are two differences with respect to a RISC CPU: modified general purpose ALU and SEU components. The rest consists of exactly the RISC CPU specification set out in Section 4, which is carefully written for the purpose, being predicated on ALU and SEU functionalities without making any requirements of them.

We denote the domain of encrypted data words for the KPU as $\mathbf{D}'$. It can be that $\mathbf{D}'$ takes up more bits than $\mathbf{D}$ (the space of unencrypted data words). One practical choice has $\mathbf{D}$ as the space of 32-bit words, and $\mathbf{D}'$ (a subspace of) 64-bit words, which fits well with block ciphers such as DES [14] and Blowfish [13]. Nevertheless, we will assume here that elements of $\mathbf{D}'$ take up the same number of bits as elements of $\mathbf{D}$, in order to avoid minor technical adjustments.

Thus a KPU is obtained by taking the ordinary RISC CPU design of Section 4 and replacing the ALU and SEU, with functionalities Λ_ι and Ξ respectively, with units that have modified functionalities

$$\Lambda'_{\iota'} : \mathbf{D}' \times \mathbf{D}' \to \mathbf{D}' \times \mathbf{D}' \qquad\qquad \Xi' : \mathbf{D}' \to \mathbf{D}'$$

that are designed to satisfy certain equational 'correctness' conditions derived from the following architectural facts about a RISC processor:

(i) the first output of the ALU drives the 32-bit arithmetic result;
(ii) the second ALU output drives the top part in 64-bit arithmetic results;
(iii) when appropriate, a non-zero second ALU output triggers a program branch;
(iv) the SEU output produces a 32-bit result from a 16-bit input.

Because of that wiring in the processor, the required relation (1) between original and modified transition functions μ, μ' specified in Section 3 in conjunction with the detailed instruction functionalities (Section 4) then determines that (17-20) below must hold. Connection (i) determines (19), (ii) forces (20) and (iii) forces (18), while (iv) forces (17). Here π_2 is the projection to the second of the pair of outputs from the ALU or modified ALU, δ_0 is the function that returns true at 0, and false otherwise, ι_0 is an ALU function code for a comparison operator, such as *less than*, ι_1 is a function code for an arithmetic operation with one significant output, such as *addition*, ι_2 is an ALU function code for an arithmetic operation with two significant outputs, such as *multiplication*, and ι'_0, ι'_1, ι'_2 are the analogues for the modified ALU.

$$\mathcal{D} \circ \Xi' = \Xi \circ \mathcal{D} \tag{17}$$

$$\delta_0 \circ \pi_2 \circ \Lambda'_{\iota'_0} = \delta_0 \circ \pi_2 \circ \Lambda_{\iota_0} \circ (\mathcal{D} \times \mathcal{D}) \tag{18}$$

$$\mathcal{D} \circ \pi_1 \circ \Lambda'_{\iota'_1} = \pi_1 \circ \Lambda_{\iota_1} \circ (\mathcal{D} \times \mathcal{D}) \tag{19}$$

$$(\mathcal{D} \times \mathcal{D}) \circ \Lambda'_{\iota'_2} = \Lambda_{\iota_2} \circ (\mathcal{D} \times \mathcal{D}) \tag{20}$$

Constructing new ALU and SEU to satisfy (17-20) makes $\mathcal{D}$ into a homomorphism from the arithmetic structure $(\mathbf{D}, \Xi, \Lambda)$ to $(\mathbf{D}', \Xi', \Lambda')$. The decryption function $\mathcal{D}$ thus represents a 'fully homomorphic cipher', by construction, with respect to the arithmetic operations in an ALU. Although fully homomorphic encryption was thought for many years to be difficult or impossible to achieve, until [5], the construction here is trivially easy to accomplish because it is the homomorphic image of the ALU and SEU that is constructed given an arbitrary encryption, not the encryption given the arithmetic.

Formalising the argument requires a little more description of an ('abstract') KPU. Let $\mathbf{X}'$ be the space of encrypted instructions, let $\mathbf{R}' = (\mathbf{D}' \sqcup \mathbf{P})^{32}$ be the space of vectors of values in the general purpose registers, and let

$$\mathbf{I}' = \mathbf{D}' \times (\mathbf{D}' \sqcup \mathbf{P}) \sqcup \mathbf{D}' \sqcup \mathbf{P} \sqcup \mathbb{1} \qquad \mathbf{O}' = (\mathbf{D}' \sqcup \mathbf{P}) \sqcup \mathbb{1}$$
$$i' = \quad a'!d', \, a'!p \quad | \; a'? \; | \uparrow p \; | \; \phi \qquad o' = \quad d', \, p \quad | \; \phi$$

respectively be the space of encrypted memory and program counter change requests emitted and the space of encrypted responses received.

What is the reason for the 'abstract' qualifier? The values in registers and memory here can be thought of as carrying an extra marker that distinguishes between encrypted data or data addresses in $\mathbf{D}'$ and unencrypted program addresses in $\mathbf{P}$. That can be seen where we write $\mathbf{D}' \sqcup \mathbf{P}$ (the disjoint union of two sets), instead of $\mathbf{D}' \cup \mathbf{P}$ (the ordinary non-disjoint union of two sets). We will need the markers in order to extend decryption $\mathcal{D}$ to $\mathbf{O}' \to \mathbf{O}$ and $\mathbf{I}' \to \mathbf{I}$ in Section 6. Afterwards, we will drop the markers and get an 'ordinary' KPU.

Definition 2. *A (abstract) KPU is a RISC CPU with the ALU and SEU changed as per (17–20). The modified transition function $\mu' : \mathbf{R}' \times \mathbf{X}' \times \mathbf{O}' \to \mathbf{R}' \times \mathbf{I}'$:*

$$(R'_{n+1}, i'_n) = \mu'(R'_n, x'_n, o'_n) \tag{10'}$$

is obtained by replacing Ξ by Ξ' and Λ by Λ' in each of the defining equations for μ per instruction in the reference RISC CPU (see Section 4).

The nth instruction $x'_n : \mathbf{X}'$ executed by the KPU is the one whose address is in the KPU's program counter $p' : \mathbf{P}$ on the nth cycle, retrieved from program memory $P' : \mathbf{P} \to \mathbf{X}'$, as stated in (9'). That is, $x'_n = P'(p'_n)$.

Memory M' and program counter p' in a KPU function exactly as memory M and program counter p do in a RISC CPU. That is, they satisfy (11–13) with $M' : \mathbf{D}' \to (\mathbf{D}' \sqcup \mathbf{P})$ substituted for M and $p' : \mathbf{P}$ for p:

$$M'_{n+1}(a') = \begin{cases} \theta & i'_n = a'!\theta \\ M'_n(a') & \text{otherwise} \end{cases} \tag{11'}$$

$$o'_n = \begin{cases} M'_n(a') & i'_n = a'? \\ \phi & \text{otherwise} \end{cases} \tag{12'}$$

$$p'_{n+1} = \begin{cases} p' & i'_n = \uparrow p' \\ p'_n + 4 & \text{otherwise} \end{cases} \tag{13'}$$

for addresses $a':\mathbf{D}'$ and possible data $\theta = d':\mathbf{D}'$, $p':\mathbf{P}$.

Certain type-safety requirements arise in order that (10') be valid. For example, the functionality for an instruction $\mathbf{addiu}' \, \rho_1 \, \rho_2 \, k'$ is as follows in a KPU, replacing Ξ by Ξ' and Λ by Λ' in the corresponding CPU equation (14):

$$R'_{n+1}(\rho) = \begin{cases} \pi_1(\Lambda'_{\text{add}}(R'_n(\rho_2), \Xi'(k'))) & 0 \neq \rho = \rho_1 \\ R'_n(\rho) & \text{otherwise} \end{cases} \tag{14'}$$

$$o'_n = i'_n = \phi$$

for $k' : \mathbf{D}'$ and $R'_n(\rho_2) : \mathbf{D}'$. The k' in **addiu'** $\rho_1\,\rho_2\,k'$ will be the encryption of some value k, with $\mathcal{D}(k') = k$. That is a compile-time type-safety requirement: data embedded in the KPU instruction must be encrypted. Moreover, $(14')$ is only valid when $R'_n(\rho_2)$ contains an encrypted value – that is, at run-time, the **addiu'** $\rho_1\,\rho_2\,k'$ instruction gets an encrypted value in the register with index ρ_2 (and puts one in the register with index ρ_1). The functionality $(15')$ for the instruction **jal'** p in a KPU is unchanged from the functionality (15) for **jal** p in a CPU. However, the functionality for an instruction **lw'** $\rho_1\,k'(\rho_2)$ is as follows in a KPU, replacing $\varXi$ by $\varXi'$ and $\varLambda$ by $\varLambda'$ in the CPU equation (16):

$$R'_{n+1}(\rho) = \begin{cases} d' & 0 \neq \rho = \rho_1 \\ R'_n(\rho) & \textbf{otherwise} \end{cases} \tag{16'}$$

$$i'_n = a'?\quad \textbf{where}\ a' = \pi_1(\varLambda'_{\mathrm{add}}(R'_n(\rho_2), \varXi'(k')))$$
$$o'_n = d'\qquad\qquad d' = M'_n(a')$$

The k' embedded in the instruction at compile-time must be the encryption of some value k, and at run-time the register with index ρ_2 must contain an encrypted value – to which k is then added, under the encryption. Other instructions generate their own type-safety requirements. See [2] for the list.

In summary, a KPU has the same design as the reference RISC CPU, but with a different ALU and SEU dropped in. As a result, different values circulate as data and data addresses. Certain type-safety requirements with respect to encrypted versus unencrypted data types are generated as a result of the swap. One might suspect that something goes wrong when encrypted values are used as indirect memory references, or in other complicated circumstances, but that is not the case: in the next section we will prove that the circulating data is not nonsense, but an encrypted version of what would circulate inside a RISC CPU.

6   Correspondence of a Running KPU to a RISC CPU

For readability, we extend the decryption function $\mathcal{D}$ to the encrypted instruction space $\mathbf{X}'$ by mapping the instruction **addiu'** $\rho_1\,\rho_2\,k'$ to **addiu** $\rho_1\,\rho_2\,\mathcal{D}(k')$, and so on. That is to say, encrypted instructions embed an encrypted opcode and encrypted embedded data, but unencrypted register indices (and program addresses), and decryption consists of mapping the opcode and embedded data, while leaving register indices and program addresses untouched.

We also extend $\mathcal{D}$ to encrypted request space $\mathbf{I}'$ and response space $\mathbf{O}'$ as follows, taking $i' : \mathbf{I}'$ to $i : \mathbf{I}$ and $o' : \mathbf{O}'$ to $o : \mathbf{O}$:

$$o = \begin{cases} \mathcal{D}(d') & o' = d' \\ p' & o' = p' \\ \phi & o' = \phi \end{cases} \qquad i = \begin{cases} \mathcal{D}(a')!\mathcal{D}(d') & i' = a'!d' \\ \mathcal{D}(a')!p' & i' = a'!p' \\ \mathcal{D}(a')? & i' = a'? \\ \uparrow p' & i' = \uparrow p' \\ \phi & i' = \phi \end{cases}$$

for $a' : \mathbf{D}'$, $d' : \mathbf{D}'$, $p' : \mathbf{P}$. I.e., unencrypted program addresses embedded in requests and responses are left untouched, and encrypted data and data addresses are decrypted. That now allows us to derive equation (1) formally:

Proposition 1. *The transition function μ' of the (abstract) KPU is a homomorphic image of the transition function μ of the reference RISC CPU:*

$$\mu \circ (\mathcal{D} \times \mathcal{D} \times \mathcal{D}) = (\mathcal{D} \times \mathcal{D}) \circ \mu' \tag{21}$$

Proof. By consideration of $\mu(-, x, -)$ and $\mu'(-, x', -)$ case-by-case for each different instruction x and its encrypted analogue x'. That requires checking the specifications given in full in [1] and partially set out in three exemplars here (14–16, 14'–16'), and agreeing that they are related via (17–20).   □

Equation (1) is (21) rewritten. Now we are in position to prove Theorem 1:

Theorem 2. *Suppose that (2) holds, i.e., $P(p) = \mathcal{D}(P'(p))$ at program addresses p of a type-safe program in P and P'. Then the following relations hold on cycle n between the register states R, memory states M, and I/O i, o of a (abstract) KPU and the reference RISC CPU, provided that they hold at cycle 0:*

$$p_n = p'_n \tag{22}$$
$$R_n = \mathcal{D} \circ R'_n \tag{23}$$
$$M_n \circ \mathcal{D} = \mathcal{D} \circ M'_n \tag{24}$$
$$o_n = \mathcal{D}(o'_n) \tag{25}$$
$$i_n = \mathcal{D}(i'_n) \tag{26}$$

Proof. By induction, given equations (10–13) and (10'–13') and the homomorphism (21). Type-safety means that each of the specifications for the different functional parts of the modified ALU within the KPU gets the kind of values that are expected (in $\mathbf{D}'$ or in $\mathbf{P}$, as appropriate), allowing (21) to hold.   □

Theorem 1 follows as a corollary, on dropping the type markers on register and memory contents to get an 'ordinary' KPU. In the latter, encrypted and unencrypted data can hypothetically be confused, but the program running is safe from that by hypothesis (type-safe, or 'crypto-safe'), so it never happens.

7   Related Work

Sagedy [12] pursues the same ideas as motivated KPU design, but does not show that codecs can be discarded or recognize the mathematics allowing it. And while the relation of the mathematics to Gentry's discovery of homomorphic encryption [5] is evident, it played no part in KPU development.

Theoretical work that may be relevant to KPU technology in future includes that on 'oblivious ram machines' [6, 7], where the sequence of memory accesses is

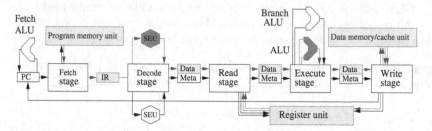

Fig. 5. Schematic of a pipelined KPU implementation. Pathways and units through which encrypted data flows are in light shading. The two combinatorial logic units which differ from the standard MIPS RISC processor units are in solid fill.

not related to the program input. In a KPU, the sequence is as deterministic as in a CPU, albeit encrypted, and one can envisage signalling using recurring access counts, for example. That becomes of interest if one is induced to run a program that has been Trojaned before encryption, in which case such signals might be used to give away the encryption. Note that KPU memory is ordinary RAM in every respect, however – it merely holds encrypted data – so the relevance is orthogonal to the particularity that a KPU is involved.

More directly relevant are hardware offerings, such as IBM's Hardware Security Module (HSM), which provides a small secure sandpit memory. Only one process has access to the sandpit, and that process excludes any other process from running at the same time (so observing it via software is not possible). When the process leaves the processor, memory is flushed to the sandpit, and recovered again at next entry. Physical probes can see the unencrypted data. A KPU, in contrast, admits multiple processes, indeed observers, with full access – a keyless design can safely be run as a software simulation. But there is nothing to prevent someone adding a HSM to a KPU design, and gaining security advantages from both, so these alternatives should not be regarded as exclusive. In particular, we may envisage a keyed KPU design with a HSM that holds the keys for an integrated combined codec and ALU/SEU (the units in solid fill in Fig. 5) in a secure area on the processor chip, thereby allowing different KPU encryptions to be dynamically configured in hardware for different processes.

Lie *et al.* have developed a hardware processor architecture [9] with memory that is divided into areas that are only accessible to processes executed from within the compartment. A virtual XOM machine is possible but the underlying hardware needs to support a unique private key, private memory, and traps on cache misses (a KPU design also benefits from modifications of this kind). For efficient operation, hardware assistance for fast symmetric cipher encryption (such as DES[3]) is also required. In other words, a XOM machine by design decrypts the encrypted instruction and data stream internally and thus has an intrinsic vulnerability to physical probes. The XOM design is 'missing the trick' of the KPU design principle – that it suffices to only change the arithmetic in the processor in order to change the encryption of all the data and data addresses at once, without doing any encryption.

The CryptoPage architecture [4] by Duc and Keryell is a whole-processor solution that has some commonalities with the KPU design presented here. Indeed, conversations between those authors and the first author of this paper took place in 2006, and the visualization in Fig. 2 is owed to those conversations. However, CryptoPage appears to stop short of the complete design transformation that produces a KPU. Instead, only the memory area in a CryptoPage processor is protected via encryption, and it is the address space that is encrypted. Data in memory may be encrypted via exogenous hardware units. The design is not intended to be proof against hardware probes, though it has the same objective as a KPU – namely, protecting the running program against unwanted observation or tampering by the operators of the computer system. The CryptoPage architecture builds on the HIDE [16] bus and cache architecture, which randomizes memory accesses via dynamic cache remapping, so that the same access repeated looks different the second time. The architecture could usefully be incorporated in a KPU, preventing a trojan program signalling to an outside observer via repeated memory accesses.

8 Conclusion

This paper has described how a KPU ('general purpose crypto-processor') may run encrypted but otherwise ordinary RISC machine code, storing encrypted data and addresses in memory and registers without any encryption or decryption necessarily taking place. When that is the case, security is not compromised if the KPU is run in simulation because there is no covert 'decrypted form' of the encrypted data for probes to uncover. The technology promises security for programs in a remote environment, protecting data and results from the operators. A KPU design is compatible with other hardware-based security solutions, which may be advantageously combined with it.

9 Future Work

We have not discussed here any details of an ALU implementation for a KPU, and we continue to investigate the security properties of many design schemes that produce the modified arithmetics required. Modular keyless solutions, which embed encrypted arithmetic without internal codecs, have readily computed cipher strengths and their design is a question of trading off the desired strength against size and complexity. The keyed hardware-only designs that we are investigating integrate the ALU with codecs that use keys kept safe in one of IBM's security modules when not in use.

References

[1] Breuer, P.T.: Encrypted Data Processing. Patent pending, UK Patent Office #GB1120531.7. UK (November 2011)

[2] Breuer, P.T., Bowen, J.P.: Typed Assembler for a RISC Crypto-Processor. In: Barthe, G., Livshits, B., Scandariato, R. (eds.) ESSoS 2012. LNCS, vol. 7159, pp. 22–29. Springer, Heidelberg (2012)

[3] Buchty, R., Heintze, N., Oliva, D.: Cryptonite – A Programmable Crypto Processor Architecture for High-Bandwidth Applications. In: Müller-Schloer, C., Ungerer, T., Bauer, B. (eds.) ARCS 2004. LNCS, vol. 2981, pp. 184–198. Springer, Heidelberg (2004)

[4] Duc, G., Keryell, R.: An Efficient Secure Architecture with Memory Encryption, Integrity and Information Leakage Protection. In: ACSAC 2006, Proceedings of the 22nd Annual Computer Security Applications Conference, Miami Beach, FL, USA, pp. 483–492. IEEE Computer Society, Washington, DC (2006), ISBN:0-7695-2716-7, doi:10.1109/ACSAC.2006.21

[5] Gentry, C.: Fully Homomorphic Encryption Using Ideal Lattices. In: Proc. 41st ACM Symposium on Theory of Computing, pp. 169–178. ACM (2009) doi: 10.1145/1536414.1536440, ISBN: 978-1-60558-506-2

[6] Goldreich, O.: Towards a theory of software protection and simulation by oblivious RAMs. In: Proc. 19th ACM Symp. on Theory of Computing, pp. 182–194. ACM (1987), doi:10.1145/28395.28416, ISBN: 0-89791-221-7

[7] Goldreich, O., Ostrovsky, R.: Software protection and simulation on oblivious RAMs. Journal of the ACM (JACM) 43(3), 431–473 (1996), doi:10.1145/233551.233553

[8] MIPS Technologies Inc. MIPS32 4K Processor Core Family Software User's Manual. MD00016. 1225 Charleston Road, Mountain View, CA 94043-1353 (January 2001)

[9] Lie, D., et al.: Architectural support for copy and tamper resistant software. ACM SIGPLAN Notices 35(11), 168–177 (2000), doi:10.1145/356989.357005

[10] Oliva, D., Buchty, R., Heintze, N.: AES and the cryptonite crypto processor. In: Proc. Intl. Conf. on Compilers, Architecture and Synthesis for Embedded Systems. ACM (2003), doi:10.1145/951710.951738

[11] Patterson, D.A.: Reduced Instruction Set Computers. Communications of the ACM 28(1), 8–21 (1985)

[12] Sagedy, C.: ECEC 490: Processor Design Project Page (December 2008), http://chris.sagedy.com/projects/ecec490_fa08/#encrypted

[13] Schneier, B.: Description of a New Variable-Length Key, 64-Bit Block Cipher (Blowfish). In: Anderson, R. (ed.) FSE 1993. LNCS, vol. 809, pp. 191–204. Springer, Heidelberg (1994)

[14] National Bureau of Standards. Data Encryption Standard. FIPS-Pub.46. U.S. Department of Commerce, Washington, D.C., USA (January 1977)

[15] Sun, M.-C., et al.: Design of a scalable RSA and ECC crypto-processor. In: Proc. ASP-DAC 2003: Asia and South Pacific Design Automation Conf. ACM (2003), doi:10.1145/1119772.1119874

[16] Zhuang, X., Zhang, T., Pande, S.: HIDE: an infrastructure for efficiently protecting information leackage on the address bus. In: Proc. 11th Intl. Conf. on Architectural Support for Programming Languages and Operating Systems (ASPLOS), pp. 72–84. ACM Press (October 2004)

DKAL*: Constructing Executable Specifications of Authorization Protocols

Jean-Baptiste Jeannin[1], Guido de Caso[2], Juan Chen[3],
Yuri Gurevich[3], Prasad Naldurg[3], and Nikhil Swamy[3]

[1] Cornell University
[2] Universidad de Buenos Aires
[3] Microsoft Research

Abstract. Many prior trust management frameworks provide authorization logics for specifying policies based on distributed trust. However, to implement a security protocol using these frameworks, one usually resorts to a general-purpose programming language. To reason about the security of the entire system, one must study not only policies in the authorization logic, but also hard-to-analyze implementation code.

This paper proposes DKAL*, a language for constructing executable specifications of authorization protocols. Protocol and policy designers can use DKAL*'s authorization logic for expressing distributed trust relationships, and its small rule-based programming language to describe the message sequence of a protocol. Importantly, many low-level details of the protocol (e.g., marshaling formats or management of state consistency) are left abstract in DKAL*, but sufficient details must be provided in order for the protocol to be executable.

We formalize the semantics of DKAL*, giving it an operational semantics and a type system. We prove various properties of DKAL*, including type soundness and a decidability property for its underlying logic. We also present an interpreter for DKAL*, mechanically verified for correctness and security. We evaluate our work experimentally on several examples.

1 Introduction

Despite many years of successful research in protocol design, federated cloud services continue to be plagued by flaws in the design and implementation of critical authorization protocols. For example, recent work by Wang et al. [24] reveals authorization errors in a variety of federated online payment services. Among other reasons, Wang et al. argue that the ad hoc implementation of such services obscures the delicate protocols on which they are based, making the design and implementation of these protocols difficult to analyze for vulnerabilities. We propose to address such difficulties by providing a domain-specific language to concisely specify authorization protocols so that the protocol design is *evident* (and suitable for security analysis) and *executable*.

To illustrate, consider the following scenario. An online retailer W wishes to use a third-party payment provider P (e.g., PayPal) to process payments. As Wang et al. report, many of the existing tools used to build such a website are often buggy, with no clear specification of the protocol they implement.

J. Jürjens, B. Livshits, and R. Scandariato (Eds.): ESSoS 2013, LNCS 7781, pp. 139–154, 2013.
© Springer-Verlag Berlin Heidelberg 2013

Informally, we would like to start by specifying that the retailer W trusts P to process payments. Prior authorization logics allow such trust relationships to be expressed concisely; e.g., in infon logic [13], one might write a policy for W stating that it is safe to conclude that a principal c paid n for order oid, if P said so: $\forall$c, oid, n. P said Paid(c, oid, n) $\Longrightarrow$ Paid(c, oid, n).

However, the means to arrive at a specific authorization protocol based on this trust relationship alone is unclear. Even a simple protocol involves several rounds of communication between a customer C, the website W, and the payment provider P. For example, the protocol in Figure 1 involves five steps: (1) a customer C requests to purchase some item i for a price n; (2) the retailer W requests C to provide a certificate from PayPal (P) authorizing C's payment; (3) C forwards the payment request to P; (4) P authorizes the payment from C to W and issues a certificate confirming the payment; (5) W, relying on a trust relationship with P, concludes that the payment has indeed been processed and ends the protocol by returning a confirmation to C.

Typically, one implements such protocols in a general-purpose programming language, where one makes queries to a trust management engine (e.g., SecPAL [3]) to determine if access to a protected resource is to be permitted. While this approach provides flexibility, it leaves the design of the authorization protocol unclear, and opens the door to vulnerabilities due to improper protocol design or other mundane programming errors. Of course, such errors can be detected by using semi-automated program verification tools, but this demands considerable expertise. Besides, even for experts, a methodology in which the protocol design is made evident by construction, facilitates simpler analysis.

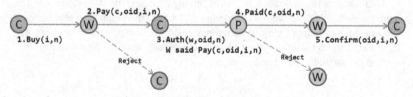

Fig. 1. A simple protocol for processing online payments

To address these problems, we propose DKAL*, a domain-specific language for executable specifications of authorization protocols. We formalize the semantics of DKAL* and implement a verified interpreter using F*, a verification-oriented dialect of ML. DKAL* programs include three conceptual components: the quantified primal infon logic (QPIL) for expressing distributed trust relationships; a small rule-based programming language for describing message flow of protocols; and finally, DKAL* programs may embed F* expressions, the host language of our interpreter—one can use this facility to evaluate arithmetic expressions, connect to databases, etc. Thus, having designed a protocol in DKAL*, one may readily obtain an executable implementation in F*. Once in F*, the source code can, in principle, be directly analyzed for high-level security properties using F*'s type system and related tools such as the Crypto Verification Toolkit [6]. However, in this work, we take DKAL* programs as specifications, and our interpreter is proven to faithfully implement the specification, regardless of any end-to-end security objective that the DKAL* programmer may have had in mind.

Figure 2 shows an example of DKAL* code, a policy specified by each of the three principals in our online retail scenario. DKAL* programs are a collection of rules, each of which can be thought of as handlers that cause specific *actions* to occur in response to events that meet certain *conditions*. Actions include sending messages (send), forwarding messages (fwd), a logging facility (log), generating fresh identifiers (with fresh), and introducing new information (learn) to the principal's QPIL knowledge base. Conditions have two forms: when e is satisfied if the principal has received a message that matches the pattern constructed by the term e; the condition if e is satisfied if the proposition constructed by the term e is derivable in QPIL. Terms include the form eval(e), where e is an F* expression evaluated by the interpreter; variables (e.g., i, n); constants (e.g., W, P); and constructed terms (Buy(i, n), etc.).

```
(* CUSTOMER's (C) policy *)
C1:
when C said Click(i, n) then
send W (C said Buy(i, n))
log (C said Init(W, i, n))


C3:
when C said Init(w, i, n)
when w said Pay(C, oid, i, n) as m1
then send P (C said Auth(w, oid, n))
     fwd P m1



(* PAYPAL's (P) policy *)
P4 :
when c said Auth(w, oid, n)
when w said Pay(c, oid, i, n)
if eval(checkBalance "c" "n") then
send w (P said Paid(c, oid, n))
where checkBalance = (* F* code *)
```

```
(* Website's (W) policy *)
W2:
when c said Buy(i, n)
if eval(checkPrice("i","n")) then
with fresh oid
send c (W said Pay(c, oid, i, n))
log (W said Pay(c, oid, i, n))
where checkPrice = (* F* code *)

W5:
when W said Pay(c,oid,i,n)
if Paid(c, oid, n) then
send c (W said Confirm(oid,i,n))


W6:
when P said x as i then
learn i

W7:
learn ∀p,oid,n.
     P said Paid(p, oid, n)
     ⟹Paid(p, oid, n)
```

Fig. 2. A DKAL* policy implementing the online retail protocol

Rules C_1, W_2, C_3, P_4, W_5 correspond to the steps (1)-(5) in Figure 1: (1) Rule C_1 initiates the protocol in response to a click issued by the customer. C sends a Buy message to the website W. C also logs an Init message to indicate that she has initiated the transaction. (2) After receiving the Buy message, W applies rule W_2 to request a payment certificate. W checks the price of the item (by calling an F* function), and sends a message Pay to C requesting payment. W also logs a message to keep track of the transaction currently underway. (3) Once C gets such a message from W and checks her log for the Init message, she applies rule C_3 to forward a payment request to P.

(4) Rule P_4 is P's policy that authorizes the payment by sending a Paid message to the website W (after checking and updating C's balance, using $F^\star$). (5) If W receives a Paid message, she uses her trust assumption in P (W_6 and W_7), and a decision procedure for QPIL to conclude that the item is paid, and sends a confirmation message to C (W_5).

This paper makes several technical contributions.

(i) We formalize the design of DKAL* and analyze the central entailment relation of QPIL. We give an operational semantics and a type system for DKAL* and prove that execution is insensitive to the order of rule evaluation. Our semantics provides the formal basis on which to analyze DKAL* policies.

(ii) We provide an interpreter for DKAL* in $F^\star$. We mechanically check with $F^\star$ that our interpreter soundly implements the formal semantics of DKAL*, including a verified implementation of a decision procedure for QPIL. Our interpreter includes a verified protocol based on public-key cryptography for establishing message authenticity, where we can mechanically check that recipients only accept authentic messages. Using refinement type checking, we show how to securely embed and evaluate $F^\star$ terms within DKAL*, allowing a DKAL* protocol to easily and safely interface with its environment.

(iii) We report on an experimental evaluation of DKAL* by developing a suite of 8 examples. Our experience indicates that DKAL* specifications can be terse, conveying the important high-level aspects of a distributed security protocol, while leaving many of the low-level details necessary to produce an executable implementation to our verified interpreter.

2   QPIL: **Quantified Primal Infon Logic**

We first review QPIL, Gurevich and Neeman's primal infon logic with quantifiers. Gurevich and Neeman introduced QPIL pragmatically, because of its combination of feasibility and expressivity. But QPIL is arguably one of two intrinsic logics of information (used by arbitrary principals for communication and reasoning) [4]. Our formulation differs from theirs in that we pay close attention to binders to facilitate a mechanically verified implementation of its decision procedure.

Syntax. QPIL has two basic concepts. The first is *infon*, a formula which represents a unit of information (which may be learned, communicated, etc.). The second is *evidence*—an infon i may be accompanied by a term t which serves as evidence for the validity of i. The form of evidence is left abstract; e.g., an infon i may be accompanied by a digital signature to serve as evidence that it was communicated by a principal p; or, it may represent a proof tree recording a derivation of i from some set of hypotheses according to the inference rules of the logic.

The syntax of QPIL is shown below. Predicates Q and constants c are subscripted with their types, although we elide the subscripts when the types are unimportant. Types include booleans and integers (and other common types), principals, and a distinguished type for evidence terms, ev. The terms include variables x, y, z and constants c (tagged) with their types. Later (Section 4) we add embedded $F^\star$ terms to the term language.

Infons i include the true infon $\top$; the application of a predicate symbol Q to a sequence of terms t; a conjunction form $i \wedge j$; an implication form $i \Rightarrow j$; the form

p said i, which is the modal operator of speech applied to an infon; and finally *justified infons*, Ev $t\ i$, which associates an evidence term t with an infon i. Note that when a principal sends Ev $t\ i$, he is merely asserting that t is evidence for i, and the receiver of the message, if he desires so, can check t. An example of an authorization is the infon: Bob said CanRead(Alice, `"file.txt"`). QPIL includes quantified infons ι, where an infon i may be preceded by a sequence of binders for universally quantified variables $\overline{x{:}\tau}$. The use of quantifiers allows for more general and flexible policies such as: $\forall$(x:prin). Bob said Trusted(x) $\Longrightarrow$ CanRead(x, `"file.txt"`). Quantified infons may also be justified by associating them with evidence using Ev $t\ \iota$. Unless explicitly mentioned, we blur the distinction between quantified infons and infons.

Syntax of QPIL

Meta-variables: x, y, z variables; $Q_{\overline{\tau}}$ predicates; c_τ constants

type	τ	$::=$ bool $\mid$ int $\mid$ prin $\mid$ ev	term	$p, t ::= x \mid c_\tau$
infon	i, j	$::= \top \mid Q_{\overline{\tau}}\,\overline{t} \mid i \wedge j \mid i \Rightarrow j$	quantified infon ι	$::= i \mid$ Ev $t\,\iota \mid \forall \overline{x{:}\tau}.i$
		$\mid p$ said $i \mid$ Ev $t\,i$	type context	$\Gamma ::= \cdot \mid x{:}\tau \mid \Gamma, \Gamma$
infon set	$M, K ::= \overline{\iota}$			

Typing. QPIL has three typing judgments (shown below): $\Gamma \vdash \iota$ for quantified infons; $\Gamma \vdash i$ for infons; and $\Gamma \vdash t : \tau$ for terms, where the typing context Γ maps variables to their types. Intuitively, $\Gamma \vdash \iota$ ensures that the variables of ι appear in Γ at suitable types. The typing judgments also rely on a well-formedness judgment for the context: we write Γ ok for an environment where no variable appears twice, and $\Gamma(x)$ for the type τ such that Γ contains $x : \tau$.

Typing terms and infons

$$\frac{\Gamma \text{ ok}}{\Gamma \vdash c_\tau : \tau} \qquad \frac{\Gamma \text{ ok}}{\Gamma \vdash x : \Gamma(x)} \qquad \frac{\Gamma \text{ ok}}{\Gamma \vdash \top} \qquad \frac{\Gamma, \overline{x{:}\tau} \vdash i}{\Gamma \vdash \forall \overline{x{:}\tau}.i} \qquad \frac{\forall i. \Gamma \vdash t_i : \tau_i}{\Gamma \vdash Q_{\overline{\tau}}\,\overline{t}}$$

$$\frac{\Gamma \vdash i \quad \Gamma \vdash j}{\Gamma \vdash i \wedge j} \qquad \frac{\Gamma \vdash i \quad \Gamma \vdash j}{\Gamma \vdash i \Rightarrow j} \qquad \frac{\Gamma \vdash p : \text{prin} \quad \Gamma \vdash i}{\Gamma \vdash p \text{ said } i} \qquad \frac{\Gamma \vdash t : \text{ev} \quad \Gamma \vdash \iota}{\Gamma \vdash \text{Ev}\,t\,\iota}$$

The typing rules for terms are straightforward—constants are typed using their subscripts, and variables by the typing context. The rules for infons are straightforward, with only one subtle point to mention. The last rule is overloaded to apply to both justified infons and justified quantified infons.

Entailment. We define an entailment relation for QPIL, a Hilbert-style calculus defining the inference rules of the logic. Our formulation relies on the notion of a prefix π, a possibly empty sequence of terms $\overline{t}$ of type prin. We write $\pi\,i$ to mean i when π is empty, or t said $(\pi'\,i)$ when $\pi = t, \pi'$. The calculus includes two relations, $K; \Gamma \vDash \iota$ for quantified infons and $K; \Gamma \vDash i$ for infons. The context in each of these relations includes an *infostrate*, K, a set of infons, representing a principal's knowledge, and a typing context Γ. We write K ok for an infostrate where for each $\iota \in K$ we have $\cdot \vdash \iota$, i.e., K is a set of well-typed closed infons. We write $K; \Gamma$ ok for (K ok and Γ ok).

Entailment relations: $K; \Gamma \vDash \iota$ **and** $K; \Gamma \vDash i$

$$\frac{K; \Gamma \text{ ok} \quad \Gamma \vdash \pi \top}{K; \Gamma \vDash \pi \top} \text{ T} \qquad \frac{K; \Gamma \text{ ok} \quad \iota \in K \quad \iota \equiv_\alpha \iota' \quad \Gamma \vdash \iota'}{K; \Gamma \vDash \iota'} \text{ Hyp-K}$$

$$\frac{K; \Gamma \vDash \pi i \quad K; \Gamma \vDash \pi j}{K; \Gamma \vDash \pi(i \wedge j)} \wedge \text{-I} \qquad \frac{K; \Gamma \vDash \pi(i \wedge j)}{K; \Gamma \vDash \pi i} \wedge \text{-E1} \qquad \frac{K; \Gamma \vDash \pi(i \wedge j)}{K; \Gamma \vDash \pi j} \wedge \text{-E2}$$

$$\frac{\Gamma \vdash \pi i \quad K; \Gamma \vDash \pi j}{K; \Gamma \vDash \pi(i \Rightarrow j)} \Rightarrow \text{-WI} \qquad \frac{\begin{array}{c} K; \Gamma \vDash \pi(i \Rightarrow j) \\ K; \Gamma \vDash \pi i \end{array}}{K; \Gamma \vDash \pi j} \Rightarrow \text{-E} \qquad \frac{K; \Gamma \vDash \pi (\text{Ev } t \, \iota)}{K; \Gamma \vDash \pi \iota} \text{ Ev-E}$$

$$\frac{K; \Gamma, \overline{x{:}\tau} \vDash i}{K; \Gamma \vDash \forall \overline{x{:}\tau}.i} \text{ Q-I} \qquad \frac{K; \Gamma \vDash \forall \overline{x{:}\tau}.j \quad \forall i. \Gamma \vdash t_i : \tau_i}{K; \Gamma \vDash j[\overline{t/x}]} \text{ Q-E}$$

The inference rule (T) allows well-typed infon $\pi \top$ to be derived from any well-formed context. The rule (Hyp-K) allows using infostrate hypotheses $\iota \in K$, but only after they have been suitably α-converted to ι', to avoid the bound names of ι' clashing with the names in the context. The premise $\Gamma \vdash \iota'$ guarantees no name clashing. The definition of alpha equivalence, $\iota \equiv_\alpha \iota'$, is standard and elided due to space constraints.

The rule ($\wedge$-I) is an introduction rule for conjunctions, with ($\wedge$-E1) and ($\wedge$-E2) the corresponding elimination rules. The modality π distributes over the conjuncts.

The rule ($\Rightarrow$-WI) is the weak introduction rule for implications, and the rule ($\Rightarrow$-E) is the usual elimination form. The weak form of implication is characteristic of primal infon logic—it allows deriving $\pi(i \Rightarrow j)$ only if πj can already be derived. This may seem pointless, except for two reasons: (1) this weak form of implication lends itself to an efficient linear-time decision procedure, at least for the propositional primal infon logic; and (2) in the case of authorization, a principal may know the conclusion πj, but may be willing to share only a weaker part $\pi (i \Rightarrow j)$ with another principal.

The rule (Ev-E) is the elimination form for evidence—note that the only way of introducing justified infons is by hypothesis or by elimination. Finally, we have (Q-I) and (Q-E) for introducing and eliminating quantifiers.

With these definitions, we can state and prove our first lemma, namely that entailment derives only well-typed infons.

Lemma 1 (Entailment is well-typed). *For all K, Γ, ι, if $K; \Gamma \vDash \iota$ then $\Gamma \vdash \iota$.*

Decidability of QPIL. There exists a complete decision procedure for QPIL entailment. Gurevich and Neeman [13] present a linear-time algorithm for the multiple derivability problem for propositional primal infon logic (PIL). It relies on a *sub-formula* property of PIL entailment, namely that the derivation $K; \cdot \vDash i$ only uses the sub-formulas of K, i. QPIL exhibits an analogous property. We refer the reader to a companion technical report for the full development [16].

While the existence of a complete decision procedure for QPIL is useful, the rest of DKAL* is designed so that it may also be used with other, more powerful authorization logics, e.g., the full infon logic with a more standard form of implication introduction.

3 The Design and Semantics of DKAL*

We now define DKAL*, a rule-based language for specifying the communication patterns in an authorization protocol. DKAL* artifacts are, simultaneously, *programs*, *policies* and *specifications*—we use the terms interchangeably, unless explicitly noted otherwise. This section introduces DKAL*'s syntax and semantics, relying on our online retail scenario for illustrative examples.

Syntax of DKAL*. The display below shows the syntax of DKAL*. A program R is a finite set of rules, each of the form $(C$ then $A)$. The semantics of DKAL* executes a program by evaluating the guards C of each rule against a principal's local configuration, and applying the actions A of only those rules whose guards are satisfied. The local configuration P of a principal p is a triple (K, M, R). It includes (1) an *infostrate*, K, which is a monotonically increasing set of infons, representing p's knowledge; (2) a *message store*, M (also a set of infons), which p may use to retain messages that it receives; and, (3) the *program* R itself. The global configuration G is the parallel composition of configurations (p, P), one for each principal p. We give a message-passing semantics for DKAL* in which the reduction of a local configuration P causes infons to be sent to other principals.

Syntax of DKAL* (with syntactic sugar on the right)

program	$R ::= C$ then $A \mid R\,R \mid \cdot$	when ι then A =	upon Ev $x\,\iota$ as m
local cfg.	$P ::= (K, M, R)$		then $(A, \text{drop }m)$
global cfg.	$G ::= (p, P) \mid G \parallel G$		for fresh x and m
guards	$C ::= \text{upon } \iota \text{ as } x \mid \text{if } \iota \mid C\,C \mid \cdot$	$\log \iota$ =	send Self ι
actions	$A ::= \text{send } p\,\iota \mid \text{fwd } p\,\iota \mid \text{drop } \iota$		
	$\mid \text{learn } \iota \mid \text{with fresh } x\,A \mid A\,A$		
infon	$i ::= \ldots \mid x$		
typing ctxt.	$\Gamma ::= \ldots \mid x\text{:infon} \mid x\text{:qinfon}$		

Guards come in two flavors. The guard (upon ι as x) is a pattern which checks whether a message matching ι is present in the principal's message store M and binds the message to x if matched. We extend the syntax of infons i so that they may contain pattern variables x. Evaluating an upon condition requires computing a substitution σ for the pattern variables such that $\sigma\,\iota$ is in the message store M. In order to ensure that pattern variables are properly used, we extend our syntax of typing environments Γ to include bindings for variables typed as infons and quantified infons (qinfon).

Guards also include boolean conditions of the form (if ι). Evaluating this guard involves a call to a decision procedure of QPIL to check that the infon ι is derivable from the principal's knowledge K. If derivable, the actions of the rule are applied; otherwise the rule is inactive. This kind of guard does not bind pattern variables.

Actions include (send $p\,\iota$), which sends ι to p authenticated by the sender; (fwd $p\,\iota$), which forwards a previously received message to p; (drop ι), which deletes a message from M; (learn ι), which adds an infon to the knowledge K; and, finally, a construct (with fresh $x\,A$) to generate fresh identifiers. In writing examples, we also use the syntactic sugar shown at the right of the display, where Self is a principal constant for the local principal.

Operational semantics of DKAL*   The operational semantics of DKAL*, deriving from
semantics of ASMs, are carefully set up to ensure a few properties. We discuss these
properties informally here, motivating various elements of the design—we formalize
these properties in the metatheory study of Section 3.

State Consistency. We desire a semantics with a consistent notion of state updates.
To achieve this, we have a message passing semantics for global configurations. But,
the reduction of each principal's configuration P is given using a big-step reduction in
which all applicable actions from the rules in P are computed atomically, with respect
to an unchanging local state. Big steps of local evaluation are interleaved with messages
being exchanged among the principals, modifying their local states.

Determinism. We aim to ensure that the semantics of a program is independent of the
order of execution of the rules in a program R. We achieve this by evaluating the set of
actions computed from a set of rules in a canonical order.

We begin by presenting the big-step evaluation of local configurations, $P \Downarrow_p A$,
where a local configuration P for a principal p evaluates to a set of actions A. The rule
(Ev) picks a rule C then A from the rule set and evaluates its guard C. Guard evaluation
produces a set of substitutions $\bar{\sigma} = \{\sigma_1, \ldots, \sigma_n\}$ of the free variables in C such that
the conditions $\sigma_i C$ are satisfied. The actions $[\![\sigma_i\ A]\!]$ are added to the actions computed
from the evaluation of the other rules in the program. Here, the function $[\![A]\!]$ interprets
a set of actions A by introducing fresh integer constants in the actions A, as required by
the (with fresh $x\ A$) construct.

The evaluation of guards is given by the function $\text{holds}_p\ K\ M\ C$, which computes a
set of substitutions. Evaluation of multiple guards involves composing the substitutions
returned by the evaluation of each guard.

Local rule evaluation: $P \Downarrow_p A$

$$\text{Ev}\frac{(K, M, (R_1, R_2)) \Downarrow_p A' \quad \text{holds}_p\ K\ M\ C = \bar{\sigma}}{(K, M, (R_1, (C \text{ then } A), R_2)) \Downarrow_p A' \cup_i [\![\sigma_i A]\!]} \qquad \text{EvEmp}\frac{}{(K, M, \cdot) \Downarrow_p \{\}}$$

$[\![\cdot]\!] : A \to A$
$[\![A]\!]$ $\qquad\qquad\qquad = A$ when (with fresh $x\ A'$) $\notin A$
$[\![A, \text{with fresh } x\ A']\!]$ $\quad = [\![A]\!], [\![A'[c_{\text{int}}/x]\!]\!]$ for c fresh

$\text{holds}_p : K \times M \times C \to 2^\sigma$
$\text{holds}_p\ K\ M\ (\text{upon } \iota \text{ as } x) = \{(\sigma, x \mapsto \sigma\iota) \mid \sigma\iota \in M \land\ \vdash \sigma\iota \land \text{dom}\,\sigma = \text{FV}(\iota)\}$
$\text{holds}_p\ K\ M\ (\text{if } \iota) \qquad = \{id \mid K; \cdot \vDash \iota\}$
$\text{holds}_p\ K\ M\ \cdot \qquad\qquad = \{id\}$
$\text{holds}_p\ K\ M\ (C_1, C_2) = \{(\sigma_2 \circ \sigma_1) \mid \sigma_1 \in \text{holds}_p\ K\ M\ C_1\ \land\ \sigma_2 \in \text{holds}_p\ K\ M\ (\sigma_1\ C_2)\}$

Evaluation of an (upon ι as x) guard returns every substitution σ such that a well-
typed message $\sigma\iota$ can be found in the store M. Our verified implementation ensures
that messages that match patterns are always properly justified, should they contain any
evidence. For (if ι), we require that the infon ι be derivable from the hypotheses in the
infostrate K. Note that, unlike for the evaluation of (upon ι as x), the semantics requires
the infon ι to be a closed term for rule evaluation to succeed.

We now define $G \longrightarrow G'$, a small-step reduction relation for global configurations.
The single rule in the semantics (GoP) picks a principal p and evaluates the rules of p to

obtain a set of actions A, and then applies these actions atomically to the configuration G. In order to ensure that the effect of applying the actions is independent of the order of evaluation of the rules, we require that all the $(\text{drop } i)$ actions in A precede all the other actions. We do this through a unary operator on actions, $\text{order}(A)$, that reorders a set of actions A according to a partial order in which all the $(\text{drop } \iota)$ actions come first.

Reduction semantics of global configurations: $G \longrightarrow G'$

$$\text{GoP} \frac{P \Downarrow_p A}{G_1 \parallel (p, P) \parallel G_2 \longrightarrow \text{app } (G_1 \parallel (p, P) \parallel G_2) \, p \, (\text{order}(A))}$$

$\text{order} : A \rightarrow A$
$\text{order}(A) = A_1, A_2 \text{ where } A_1 = \{\text{drop } \iota | \text{drop } \iota \in A\} \text{ and } A_2 = A \setminus A_1$
$\text{app} : G \rightarrow p \rightarrow A \rightarrow G$
$\text{app } G \, p \, \cdot \qquad\qquad = G$
$\text{app } (G_1 \parallel G \parallel G_2) \, p \, A = G_1 \parallel (\text{app1 } G \, p \, A) \parallel G_2$
$\text{app } G \, p \, (A, A') \qquad = \text{let } G' = \text{app } G \, p \, A \text{ in let } G'' = \text{app } G' \, p \, A' \text{ in } G''$
$\text{app1} : (p \times P) \rightarrow p \rightarrow A \rightarrow (p \times P)$
$\text{app1 } (p, (K, M, R)) \, p \, (\text{drop } \iota) \quad = (p, (K, (M \setminus \{\iota\}), R))$
$\text{app1 } (p, (K, M, R)) \, p \, (\text{learn } \iota) \quad = (p, ((K, \iota), M, R))$
$\text{app1 } (q', (K, M, R)) \, p \, (\text{fwd } q \, \iota) \, = (q', (K, (M, \iota), R))$
$\text{app1 } (q', (K, M, R)) \, p \, (\text{send } q \, \iota) = (q', (K, (M, \text{Ev } t \, \iota), R))$

The definition of $\text{app}(G, p, A)$ applies a set of actions A according to this partial order. We use $\text{app1}(G, p, A)$ in the base cases to apply a single action, following the syntax given in section 3. Note that, when p sends or forwards a message and to model the network imperfectness, the actual recipient q' may not be the intended principal q.

A type system for DKAL*. We provide a type system to ensure that the reduction of DKAL* programs is well-behaved, i.e., that configurations remain well-typed as reduction proceeds, and that that rule evaluation is deterministic.

Arbitrary DKAL* programs may execute in undesirable ways. For example, an ill-scoped program may inject ill-typed infons into the infostrate, potentially allowing nonsensical terms to become derivable. Consider the example program upon $(\forall(\text{p:principal}). \text{ALICE said x})$ as m then learn x. When evaluating this program against a message store M that contains the infon $\forall(\text{p:principal}). \text{ALICE said Good(p)}$, the upon condition is satisfiable, with $\sigma = [x \mapsto \text{Good(p)}]$. However, applying the action $\sigma(\text{learn x})$ results in adding the term Good(p) to the infostrate, which is clearly ill-formed—the variable p has escaped its scope.

Our type system is designed to rule out this and other undesirable behaviors. After defining several judgments on rules, actions and guards, it defines a judgment G ok for well-formedness of a global configuration G. Space constraints prevent us from presenting the full details of the type system here—the companion technical report contains the full development [16].

Theorem 1 ensures that well-formedness of a configuration is preserved under reduction. The corresponding progress property (that a well-formed configuration can always make a step) is trivial, since identity steps $(G \longrightarrow G)$ are always possible. Theorem 2 ensures that the order of evaluation of rules in a local configuration does not matter.

Theorem 1 (Type soundness). *Given a configuration G such that G ok, if $G \longrightarrow G'$ then G' ok.*

Theorem 2 (Determinism of local rule evaluation). *Given a configuration G, a local configuration (p, P) such that $G \parallel (p, P)$ ok and A_1, A_2 such that $P \Downarrow_p A_1$ and $P \Downarrow_p A_2$; then* $\mathsf{app}((G \parallel (p, P)), p, A_1) = \mathsf{app}((G \parallel (p, P)), p, A_2)$.

4   A Verified Interpreter for DKAL*

This section describes our verified interpreter for DKAL*, implemented in F*, a variant of ML with a more expressive type system. F* allows programmers to write down precise specifications using dependent types where types depend on values. F*'s type checker makes use of an SMT solver to automatically discharge proofs of these specifications. F* enables general-purpose programming, with recursion and effects; it has libraries for concurrency, networking, cryptography, and interoperability with other .NET languages. After typechecking, F* is compiled to .NET bytecode, with runtime support for proof-carrying code.

We present selected elements of the mostly ML-like code of our interpreter (slightly simplified for the paper), discussing F*-specific constructs as they arise. We refer the reader to Swamy et al. [22] for full definition of F*. The full code of our verified interpreter is available from http://dkal.codeplex.com.

We highlight three key elements of our interpreter:

A Verified Decision Procedure for QPIL. We formalize the QPIL entailment relation using a collection of inductive types in F*. We then implement a unification-based, backwards chaining decision procedure for QPIL and prove it sound, i.e., that it only constructs valid entailments.

Authenticity of Infons. Whereas the previous sections left the evidence terms associated with an infon abstract, in our interpreter evidence terms are represented as digital signatures. By relying on previously developed verified libraries for cryptography, we prove a correspondence property on execution traces of DKAL* configurations.

Secure Embedding of F* *in* DKAL*. We show how to securely implement the (eval e) construct, where the term e is an F* expression embedded within DKAL*. By relying on the type checker of F*, we show that embedded terms can safely be executed without breaking the invariants of the rest of the interpreter. This mechanism significantly broadens the scope of DKAL*, empowering programmers with a powerful general-purpose programming language when needed, and allowing a DKAL* protocol to seamlessly integrate within the context of a larger secure system.

As is usual in ML, our interpreter defines DKAL* syntax using a collection of algebraic types. We separate the syntax of quantified infons (polyterm) from infons, but, unlike in Section 2, we use a single type term to represent both terms t and infons i. This representation is flexible in that it allows terms and infons to be represented by a single type term, but it allows malformed terms to be constructed. We recover well-formedness by expressing the typing judgment for QPIL using inductive types (see the companion technical report [16]).

Verifying a decision procedure for QPIL *entailment.* We show below our mechanical formalization of QPIL entailment and the implementation of its decision procedure. We define two mutually recursive inductive types, entails and polyentails. The type entails K G i corresponds to the judgment $K; \Gamma \models i$, and polyentails K G i corresponds to the judgment $K; \Gamma \models \iota$ (from Section 2).

```
type prefix = list term
logic function Prefix : prefix → term → term
assume ∀i. (Prefix [] i) = i
assume ∀p pi i. (Prefix (p::pi) i) = (Prefix pi (App SaidInfon [p; i]))
```

```
type entails ::                             and polyentails ::
  infostrate ⇒ vars ⇒ term ⇒ P =              infostrate ⇒ vars ⇒ polyterm ⇒ P =
| Entails_And_Elim1: K:infostrate → G:vars   | Entails_Hyp_Knowledge :
    → i:term → j:term → pi:prefix                K:infostrate → G:vars → okCtx K G
    → entails K G                                → i:polyterm{In i K} → i':polyterm
       (Prefix pi (App AndInfon [i; j]))         → alphaEquiv i i' → polytyping G i'
    → entails K G (Prefix pi j)                  → polyentails K G i'
| ...                                        | ...
```

The code above illustrates two features of F*. First, we define the notion of an infon i with a quotation prefix π (written π i in Section 2). A quotation prefix is simply a list of terms and we define a function symbol Prefix to attach a prefix to term. This function is axiomatized by the **assume** equations, allowing the SMT solver underlying F*'s typechecker to reason about applications of the Prefix function symbol. Using this construct, we can define the constructor Entails_And_Elim1, which corresponds to the rule (∧-E1).

The constructor Entails_Hyp_Knowledge corresponds to the rule (Hyp-K), with the relation okCtx representing the well-formedness of the context and alphaEquiv corresponding to the relation $\equiv_\alpha$. The premise $\iota \in K$ from (Hyp-K) is represented by the *ghost refinement* type i:polyterm{In i K}, another feature of F*. This is the type of a polyterm i for which the property In i K is derivable by the SMT solver, without the programmer to supply a (lengthy) constructive proof.

With the above types as our specification, we implement and prove sound a unification-based, goal-directed proof search procedure to (partially) decide QPIL entailment. Our algorithm is implemented by the function derivePoly, whose signature is shown below. The type says that in an infostrate K, given a quantified infon goal with free variables included in the set U, if successful in proving the goal, the function returns a substitution s whose domain includes the variables in U such that the substitution s applied to the goal is derivable from K.

```
val derivePoly: K:infostrate → U:vars → goal:polyterm
    → option (s:substitution{Includes U (Domain s)} * polyentails K [] (PolySubst goal s))
```

The completeness of QPIL comes from [7]. We aim to extend our implementation to include a complete algorithm.

Main interpreter loop. The top-level of our interpreter is the infinite loop shown in the code below. At a high level, given a program represented by a list of rules rs, the interpreter computes and applies all enabled actions, and then, unless the actions cause a change to the local state, blocks waiting for new messages before looping.

```
let rec run (rs:list rule) = let actions = allEnabledActions rs in
  let stateChanged = applyAllActions actions in
  if stateChanged then run rs else (block_until_messages_received(); run rs)
```

Conceptually, the function allEnabledActions implements the local rule evaluation judgments $P \Downarrow_p A$, while applyAllActions implements message dispatch over the network, corresponding to the global transition step in the semantics of Section 3. Recall that in our semantics the local configuration of a process, in addition to the rule set, involves two components: the infostrate K and the message store M. We represent each of these using mutable state and globally scoped references. Each interpreter also has a single global constant, me:principal, the name of principal on whose behalf the interpreter runs.

We also axiomatize rules corresponding to the holds function of Section 3, and prove that the interpreter can apply only actions that have satisfiable guard conditions. As such, we prove a soundness property for our interpreter—the set of actions executed by the interpreter is a subset of the actions that may be executed in the operational semantics of Section 3. A limitation, as in the case of the decision procedure, is that we do not prove completeness of our interpreter, i.e., we do not prove that *all* enabled actions are indeed computed and applied.

Authenticity of communications. As discussed earlier, the semantics of DKAL* presented in Section 3 is clearly insecure—a principal p can freely forge an infon. However, our setup hints at a solution: justified infons, terms of the form Ev t i carry evidence terms t that can be used to convince a recipient of the authenticity of the infon. In this section, we instantiate t using digital signatures.

Our goal is to prove an authenticity property by analyzing execution traces of a DKAL* protocol running in the presence of a Dolev-Yao network adversary. Informally, we relate an event recording the receipt of a message Ev t (q said ι) by an honest participant p at step k in an execution trace, to a corresponding event at step $k' < k$ recording the sending of the message Ev t (q said ι) by q, unless the signing key of q has been compromised, i.e., a standard correspondence property on traces [25] to establish the authenticity of communications.

We set up the verification of this property following a methodology due to Gordon and Jeffrey [12], and later in RCF [5] and F*. The basic idea is to augment the dynamic semantics of the programming language with a facility to accumulate protocol events in an abstract log, and to prove trace properties by analyzing the abstract log.

Broadly, we record the sending of messages by adding an event (Sent p i) to the log when p sends a message i, and when receiving a message, through the use of a verified library of cryptographic primitives, we attempt to prove that the corresponding Sent event is in the log, unless the key of p has been leaked to the attacker.

We give a flavor of the main elements in our proof in the companion technical report [16]—the constructions are essentially standard; the reader may consult Swamy et al. [22] for more details about our cryptographic libraries.

Embedding F* *in* DKAL*. Our interpreter provides a simple and elegant solution to extend DKAL* with more general-purpose programming facilities. The example in Section 1 embeds an F* expression checkBalance "c" "n" within a DKAL* protocol using the eval construct. When evaluating the if-condition, the interpreter executes the eval'd term by calling the F* function checkBalance defined along with the policy. Once in F*, we have the power of a full-fledged programming language at our disposal—we query a database to check if the customer has sufficient funds, update the database, and return the result (an infon) to the eval context.

Of course, one may be concerned that eval'ing an arbitrary F* term may be dangerous, e.g., it may inappropriately access internal data structures of the interpreter, or it could accept improperly signed messages, etc. However, because the eval'd term is statically typed by F*, we ensure that it never breaks any such critical invariants.

When evaluating the F* function, the interpreter passes in a variable environment as an argument, which contains bindings for each of the pattern variables in scope at the point where the eval'd term is defined. In the future, we plan to exploit this idiom at a larger scale, aiming to build and deploy full-fledged cloud services using this DKAL*/F* hybrid language.

Experimental evaluation. The table below shows 8 examples we developed using DKAL*. Configuration files contain cryptography keys and communication ports for principals. Each principal stores her policies in a DKAL* file. The DKAL* file is compiled to F* for the interpreter to evaluate the rules. We measure the sizes of configuration files (column Config), the DKAL* files (column DKAL), and the resulting F* files (column F*). All numbers are line counts of files.

Name	Description	Config	DKAL	F*
Hello world	Two parties exchange hello messages.	13	14	45
Ping-Pong	Two parties bounce messages.	13	10	54
File system	A system restricting file accesses.	15	18	89
Calculator	Integer arithmetic.	27	27	115
Turing Machine	A simulator of Turing machines.	22	40	121
Rumors	Four principals spread messages.	32	22	144
Retail	Our online retail example in the Intro.	25	59	195
Clinical Trials	Checking that a physician can conduct a trial.	57	86	296

These examples cover diverse scenarios, ranging from simple message exchanges, to authorization, arithmetic, simulating turing machines, and online retailing. "Hello world" and "Pingpong" are simple message exchanges. "File system" has user U send a justified message U said Ask("f.txt", U, "read") to the file system to request file access and responds if authorized. "Calculator" implements integer arithmetic, demonstrating the eval construct. "Turing Machine" simulates turing machines. It uses DKAL* policies to control state transitions. "Rumors" involves trust management among four parties. "Retail" is our example in Section 1.

Our most complex example is "Clinical Trials", which simulates a pharmaceutical company hiring an independent research organization to conduct a clinical trial before releasing a new drug. This scenario was originally discussed by Blass et al.[9]. Briefly, the research organization hires sites such as hospitals or labs to execute the trial. Each site finds appropriate patients and assigns physicians to work with them. We use DKAL* to specify a protocol to enforce patient privacy: patient records are guarded by a key manager, which gives only authorized physicians the keys to access patient records. The protocol involves four message exchanges among the principals and reasoning about integer arithmetic and authorization delegation. We tie the abstract DKAL* specification to a concrete implementation of messages and wire formats through the use of standard cryptographic protocols for authentication, but with implementations that are verified (in a symbolic model) against the abstract DKAL* specifications. The arithmetic reasoning is performed by embedded F* expressions.

5   Related Work

The design of DKAL* is informed by a long line of work on abstract state machines (ASMs), also called evolving algebras or dynamic structures [14], and especially by the work on applications of the specification language AsmL [15] and the ASM-based Spec Explorer tool [10]. More directly, DKAL* derives from its predecessor Evidential DKAL [9]. Evidential DKAL extends the authorization logic DKAL [13] with a construct similar to our Ev t ι. The evidential nature of DKAL is related to Necula's proof-carrying code [20] that was followed by proof-carrying authentication [2] and more recently by evidence-based audit [23] and code-carrying authorization [19].

Our work improves on Evidential DKAL in a number of ways. First, we formulate QPIL in a manner suitable for mechanical verification—the prior formulation is informal in its treatment of quantifiers and variables. Next, although Evidential DKAL suggests incorporating an ASM-based language, it does not formalize this language—our semantics is novel. Our verified implementation and embedding of F* in DKAL* is new. In the process of our verification, we found and fixed several bugs in the prior formulation, including one serious bug related to ill-scoped variables.

Our authorization logic QPIL is related to many prior logics used in a variety of trust management systems. These are too numerous to discuss exhaustively here—Chapin et al. [11] provide a useful survey. One representative however is SD3 [18], where the problem of deciding authorization by means of solving a query on a distributed database is studied. SD3 has a certified evaluator, which is related to our verified decision procedure for QPIL. Both systems not only decide the validity of a query, but also construct a proof witness. SD3 requires an additional proof checking step, whereas our system statically guarantees that we construct only valid proofs.

Another line of related work includes programming languages that are combined with authorization logics. For example, Aura [17] is a dependently typed functional programming language whose type system embeds the authorization logic DCC [1]. Aura programmers build constructive proofs of authorization before performing security sensitive operations, whereas we provide a decision procedure within the runtime, and allow the embedding of F* terms in the specification.

Compiling DKAL* to F* allows the possibility of using F*'s verification-oriented type system to prove various properties of the protocol implementation. Thus DKAL* stands to benefit both from the extensive study of properties of abstract protocol models and the automated verification of protocol implementations. This line of work, too extensive to discuss in detail here, is covered thoroughly by a recent survey on protocol verification [8].

Our approach to embedding F* terms inside DKAL* compiling the result to F* for interpretation is a weak form of meta-programming. It is related to template Haskell [21] in that after code generation, we typecheck the resulting program as a normal F* program before interpretation. However, unlike template Haskell, we do not support execution of embedded F* code when generating F* from DKAL*. As such our approach is similar to inlining assembly instructions by many C compiler, with additional type-checking before execution.

Conclusions. DKAL* is a language that allows for the specification and execution of distributed authorization protocols.We have formalized DKAL*, giving it an operational semantics and a type system. We have also built a DKAL* interpreter, mechanically verified to soundly implement its semantics. Protocol designers can use our formalization to describe and analyze their authorization policies, while programmers can use our verified interpreter to deploy them.

References

1. Abadi, M.: Access control in a core calculus of dependency. SIGPLAN Not. 41(9), 263–273 (2006)
2. Appel, A., Felten, E.: Proof-carrying authentication. In: CCS 1999, pp. 52–62. ACM (1999)
3. Becker, M., Fournet, C., Gordon, A.: SecPAL: Design and semantics of a decentralized authorization language. Journal of Computer Security 18(4), 619–665 (2010)
4. Beklemishev, L., Blass, A., Gurevich, Y.: What is the logic of information? (in preparation, 2012)
5. Bengtson, J., Bhargavan, K., Fournet, C., Gordon, A.D., Maffeis, S.: Refinement types for secure implementations. In: CSF (2008)
6. Bhargavan, K., Fournet, C., Corin, R., Zalinescu, E.: Cryptographically verified implementations for TLS. In: ACM Conference on Computer and Communications Security, pp. 459–468 (2008)
7. Bjørner, N., de Caso, G., Gurevich, Y.: From Primal Infon Logic with Individual Variables to Datalog. In: Erdem, E., Lee, J., Lierler, Y., Pearce, D. (eds.) Correct Reasoning. LNCS, vol. 7265, pp. 72–86. Springer, Heidelberg (2012)
8. Blanchet, B.: Security Protocol Verification: Symbolic and Computational Models. In: Degano, P., Guttman, J.D. (eds.) POST 2012. LNCS, vol. 7215, pp. 3–29. Springer, Heidelberg (2012)
9. Blass, A., Gurevich, Y., Moskal, M., Neeman, I.: Evidential authorization. In: Nanz, S. (ed.) The Future of Software Engineering, pp. 73–99. Springer (2011)
10. CACM Staff: Microsoft's protocol documentation program: interoperability testing at scale. Commun. ACM 54(7), 51–57 (2011),
http://doi.acm.org/10.1145/1965724.1965741
11. Chapin, P., Skalka, C., Wang, X.: Authorization in trust management: Features and foundations. ACM Computing Surveys (CSUR) 40(3), 9 (2008)
12. Gordon, A.D., Jeffrey, A.: Typing correspondence assertions for communication protocols. Theor. Comput. Sci. 300(1-3), 379–409 (2003)
13. Gurevich, Y., Neeman, I.: DKAL: Distributed-knowledge authorization language. In: 21st IEEE Computer Security Foundations Symposium, pp. 149–162. IEEE (2008)

14. Gurevich, Y.: Evolving algebra 1993: Lipari guide. Specification and Validation Methods (1995)
15. Gurevich, Y., Rossman, B., Schulte, W.: Semantic essence of AsmL. Theor. Comput. Sci. 343(3), 370–412 (2005)
16. Jeannin, J.B., de Caso, G., Chen, J., Gurevich, Y., Naldurg, P., Swamy, N.: DKAL*: Constructing executable specifications of authorization Protocols (extended version). Tech. rep., Microsoft Research (2012), http://research.microsoft.com/fstar
17. Jia, L., Vaughan, J., Mazurak, K., Zhao, J., Zarko, L., Schorr, J., Zdancewic, S.: Aura: A programming language for authorization and audit. In: ICFP (2008)
18. Jim, T.: SD3: A trust management system with certified evaluation. In: Proceedings of the 2001 IEEE Symposium on Security and Privacy, S&P 2001, pp. 106–115. IEEE (2001)
19. Maffeis, S., Abadi, M., Fournet, C., Gordon, A.D.: Code-Carrying Authorization. In: Jajodia, S., Lopez, J. (eds.) ESORICS 2008. LNCS, vol. 5283, pp. 563–579. Springer, Heidelberg (2008)
20. Necula, G.C.: Proof-carrying code. In: POPL 1997: Proceedings of the 24th ACM SIGPLAN-SIGACT Symposium on Principles of Programming Languages, pp. 106–119. ACM Press, New York (1997)
21. Sheard, T., Jones, S.P.: Template meta-programming for Haskell. In: Proceedings of the 2002 ACM SIGPLAN Workshop on Haskell, Haskell 2002. ACM (2002)
22. Swamy, N., Chen, J., Fournet, C., Strub, P.Y., Bhargavan, K., Yang, J.: Secure distributed programming with value-dependent types. In: ICFP, pp. 266–278 (2011)
23. Vaughan, J., Jia, L., Mazurak, K., Zdancewic, S.: Evidence-based audit. In: CSF 2008, pp. 163–176. IEEE (2008)
24. Wang, R., Chen, S., Wang, X., Qadeer, S.: How to shop for free online - security analysis of cashier-as-a-service based web stores. In: IEEE Symposium on Security and Privacy, pp. 465–480 (2011)
25. Woo, T.Y.C., Lam, S.S.: A semantic model for authentication protocols. In: Proceedings of the 1993 IEEE Symposium on Security and Privacy, SP 1993, pp. 178–194. IEEE Computer Society, Washington, DC (1993), http://dl.acm.org/citation.cfm?id=882489.884188

A Formal Approach for Inspecting Privacy and Trust in Advanced Electronic Services

Koen Decroix[1], Jorn Lapon[1], Bart De Decker[2], and Vincent Naessens[1]

[1] Katholieke Hogeschool Sint-Lieven, Department of Industrial Engineering
Gebroeders Desmetstraat 1, 9000 Ghent, Belgium
`firstname.lastname@kahosl.be`
[2] KU Leuven, iMinds-DistriNet,
Celestijnenlaan 200A, 3001 Heverlee, Belgium
`firstname.lastname@cs.kuleuven.be`

Abstract. Advanced information processing technologies are often applied to large profiles and result in detailed behavior analysis. Moreover, under the pretext of increased personalization and strong accountability, organizations exchange information to compile even larger profiles. However, the user is unaware about the amount and type of personal data kept in profiles, partially due to advanced interactions between multiple organizations during service consumption.

In this paper, a formal approach to inspect privacy and trust in advanced electronic services is presented. It allows to express access and privacy policies of service providers. Also, the privacy properties of multiple authentication technologies are formally modeled. From this, meaningful privacy properties can be extracted based on varying trust assumptions. Feedback is rendered through automated reasoning, useful for both users and system designers. To demonstrate its practicability, the approach is applied to the design of a travel reservation system.

Keywords: privacy, trust, electronic services, modeling.

1 Introduction

Electronic services evolve from straightforward interactions between a user and a service provider, to complex web services in which multiple organizations are involved. For instance, even in a simple e-shop, multiple organizations are involved, namely the online shop, a bank (i.e., an organization that handles payments) and a delivery service (i.e., an organization that delivers the ordered items). Personal attributes are released to each of these service providers, and users are unaware of the profiles that are kept by each organization. Although organizations should only request the attributes that are required to offer the personalized service, often much more information is collected. First, during authentication, much more information is usually released than strictly necessary. For instance, if an organization only accepts adults whose date of birth is certified in an X.509 certificate, all attributes in that certificate are exposed to the service provider. At least the public key is unique which implies that all transactions by the same user are

J. Jürjens, B. Livshits, and R. Scandariato (Eds.): ESSoS 2013, LNCS 7781, pp. 155–170, 2013.
© Springer-Verlag Berlin Heidelberg 2013

linkable. Moreover, many certificates include uniquely identifying information. Second, some service providers may collude to extend profiles indirectly. Based on information gathered by other organizations, users can be discriminated. For instance, higher prices can be proposed to rich people. Similarly, poor people can be excluded. Note further that not all organizations are equally trustworthy.

In response to this, collection, traffic and processing of data is subject to privacy legislation. The US privacy act of 1974 is a legal framework that defines the principles of *fair information practices*. One of the principles is *openness and transparency*. It means that individuals must know what data is collected and for what purpose. The European Union Directive 95/94/EC extends the US privacy act with the principle of *explicit consent*. Organizations realize this through their *privacy policies* that users must acknowledge beforehand. These legal frameworks provide the principles to which systems must adhere. In accordance with this legislation, privacy requirements are gathered that must be satisfied by the realized system. General principles [10] can be applied during system design to fulfill these rules. Organizations also enforce their privacy guidelines for system design [4]. Some of them deploy decision support tools that assist developers during system design [14]. General design principles only guide designers during system design. Unfortunately, they are inadequate to analyze systems and extract conclusions to users and designers about the privacy-friendliness. Therefore, other approaches are required.

Contribution. This paper presents a formal approach to model attributes that are collected, stored and possibly shared by organizations. Fulfilling access control conditions and service interactions lead to the release of personal information to other organizations that are not always equally trusted. Our approach formally models different types of policies supported by service providers, namely (a) access control policies, (b) storage policies, (c) distribution policies and (d) output policies. Also, our approach formally models the privacy properties of currently available credential systems. Based on the policies and credential technology used, the profiles that can be compiled by each organization can be composed with alternative trust assumptions. More specifically, our approach enables to extract (automatically) different types of feedback, that are useful for both users and system designers. For instance, based on different trust assumptions by stakeholders, the impact on profiles is automatically derived. Similarly, the approach allows to evaluate the impact of alternative credential technologies on the data in profiles.

The rest of this paper is structured as follows. An overview of existing privacy modeling approaches is presented in Section 2. Section 3 gives a general overview of the approach. Next, the approach is applied to the design of a travel reservation system in Section 4. Section 5 evaluates the approach. The paper ends with conclusions.

2   Related Work

Analyzing the privacy of a system is not an easy task. Diaz et al. [8] use the cardinality of the anonymity set as a measure of the degree of anonymity, based on information theoretic principles. Serjantov et al. [16] propose some modifications to this approach. They apply a probability distribution on the possible origin or destination of a given message, instead of anonymity sets. These approaches present a theoretical framework for analyzing existing systems, but tend to be difficult to apply in practice. Moreover, there is no direct support for the design of new systems.

Barth et al. [1] propose a framework to express privacy policies at corporate and legislation level using a temporal logic. Their approach can also be used to verify the compliance of different – maybe conflicting – policies. However, they focus on the impact of temporal properties on privacy. Tschantz et al. [18] present a formal approach to reason about the acceptibility of data collection. More specifically, they inspect if personal information that is collected, stored and/or (eventually) distributed really serves certain goals. For instance, they inspect if data that is collected by a doctor really serves the patient's treatment. In contrast, our approach more focuses on the profiles that can be built when personal information is released. Hence, both approaches are complementary. Other approaches focus on the design of privacy-friendly applications, such as approaches that are based on the *privacy by design* principle [4,14]. These incorporate elements that allow to take privacy into account from the very early stage in the design process. Such approaches situate at the project management level and only specify the design process and not the system that is designed. To design the system itself, a lower level approach is required. This can be part of a general design process.

Sindre et al. [17] use misuse cases to derive security and privacy requirements. Deng et al. [7] present a privacy threat analysis framework to fulfill privacy requirements. The authors also apply misuse cases but with a larger focus on privacy, while in the former, privacy is less prominent and considered as one of the security objectives. A risk-assessment technique prioritizes the privacy threats. These threats determine the privacy requirements and the privacy objectives to pursue.

Naessens et al. [12] present a methodology for designing controlled anonymous applications. Their methodology not only takes the privacy concerns of the user into account, but also the concerns of other stakeholders, such as accountability and the personalization of services. The methodology allows to express anonymity and control requirements at a very high abstraction level. Anonymity properties can be derived from a conceptual model. Their approach enables also the semi-automatic selection of privacy enhancing technologies to realize the objectives. Finally, resolution strategies to resolve requirement conflicts are proposed as well. However, the approach lacks formal support. Moreover, advanced services in which multiple organizations are involved cannot easily be modeled.

Besides the application layer, privacy also involves lower communication layers. Veeningen et al. [19] present a deductive system to analyze privacy at the protocol level. They represent data in a three-layer model that includes the messages that are exchanged, their content, and their context. Such approaches enable a very detailed analysis. However, it requires model concepts that are not intuitive. Furthermore, at this level it is difficult to model relational aspects (e.g., trust) between stakeholders.

3   General Overview of the Approach

Figure 1 gives an overview of the major concepts when modeling a particular system (or application). Each system consists of a user and a set of *organizations*. An organization can be a credential issuer and/or a service provider. A service provider offers one or more services that can be consumed by individuals, possibly after disclosing personal *attributes*. The attributes that are released can either be non-asserted or asserted by a trusted organization (i.e., the credential issuer). For instance, attributes that are filled in by the user in a registration form are non-asserted. In contrast, credentials can be used to prove certain personal properties. Based on the attributes that are collected and linked, pseudonymous or identifiable profiles can be compiled. A *policy* is assigned to each service. It consists of (a) the attributes that must be released (or proved) before the service can be accessed (i.e., the access policy), (b) the set of data that is stored by the service provider (i.e., the storage policy), (c) the set of data that is released to other organizations (i.e., the distribution policy) and (d) the output policy (e.g., issued credentials). This information can typically be extracted from the privacy policies that are currently used by many electronic services.

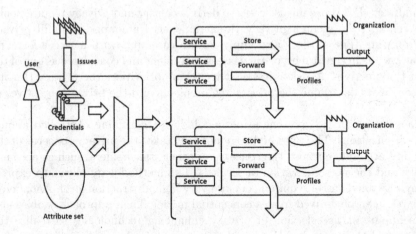

Fig. 1. Modeling electronic services

Next to the application itself, authentication technologies and trust relations are formally modeled. A formal representation of different *authentication technologies* is developed. The formal modeling focuses on the attributes that are released and linkabilities that are introduced when using a specific credential technology for fulfilling the access policy. For instance, if a user must only prove to be older than 18, only that fact is disclosed when using Idemix credential technology. However, much more info is released when using an X.509 certificate that contains additional personal attributes. Not only the exact date of birth, but also all the other attributes are visible to the service provider. Finally, each user can assign a *level of trust* to each organization involved in the system. For instance, highly trusted organizations are obeying their privacy policies, whereas untrusted organizations may share (or sell) the collected profiles with other organizations (although this is not always mentioned in their privacy policy).

The model facilitates automated reasoning using first-order logic. Note that the *IDP system* [20] was selected for the realization. It is a knowledge-based tool that allows to find solutions that fulfill a specific system specification. IDP is a declarative language that extends typed first-order logic allowing inductive definitions [6]. The latter improves the expressiveness and simplifies modeling [11] compared to classic first-order logic. IDP is used to automatically extract different types of feedback about the user's privacy that are useful for different stakeholders, such as the user and the system designer. The rest of this section is structured as follows. First, we focus on the type of feedback that can be extracted from a system specification for designers and end-users. Next, the basic concepts are formally defined after which an overview is given of the types of credential technologies and policies that are supported.

Automated Feedback Generation. The approach supports different types of feedback relevant for in-depth inspection of the user's privacy. The feedback can be classified according to three classes:

- **Information Spreading** reflects the amount and type of information that is released towards each organization. Organizations gather information during electronic service consumption and store it in profiles. A profile keeps information that is linked to the same individual. Profiles can eventually be merged.
- **Organization Behavior** shows the impact of collaborating organizations on the user's privacy. Collaboration can lead to discrimination. For instance, an insurance company can discriminate patients based on information obtained from a commercial e-health provider. However, collaborations may also have benefits for the user. For instance, data from a social network profile can be forwarded to ease registration procedures (this improves the user-experience). Changes in the business landscape, such as two organizations that merge, also influence the user's privacy (e.g., their profiles are also merged).
- **User Behavior** feedback depicts the impact of the decisions/strategies applied by the user on his privacy and trust. For instance, two e-shops can offer the same e-book at the same price but can apply different privacy policies. Similarly, using an anonymous e-cash system for electronic payments is often more privacy-friendly than using a debit card. Sometimes, multiple alternatives for authentication are possible. For instance, proving to be a resident of Brussels using an anonymous credential is more privacy-friendly than using an identity card (with X.509 certificate technology).

Furthermore, the approach allows to detect violations against ruling policies, or to detect conflicts between policies at different levels (e.g., corporate level policies conflict with governmental policies) or between different stakeholders (e.g., user policies conflict with service policies). Also, changes in policies at different levels (e.g., governmental, corporate, service specific) are covered by this approach. The model's formal policy rules can be adapted in accordance to the changes in policies.

Basic Terminology. E defines the set of stakeholders. E consists of the *user* $u \in U \subset E$ and multiple *organizations* $o \in O \subset E$. Each organization can offer a set of services, and multiple organizations can offer the same service. Therefore, $\Sigma \subseteq S \times O$ represents the set of all service instances, with S the set of services. $a \in A$ defines an attribute, with A the set of all attributes related to the user. Note that the actual attribute values are abstracted and only reflect the type of information they contain. For instance, the attribute *eye color* represents data of type *eye color*. Some attributes refer to characteristics of the user (e.g., the user's *first name*, *second name*, *date of birth*, *eye color*, and a *pseudonym* of the user's favorite chat application), while others are related to technological identifiers or parameters in the system (e.g., an *IP-address*) or environmental context (e.g., *the location*).

Organizations can store the attributes that are collected during service consumption in a profile, bundling related attributes. For instance, a profile can contain attributes revealed during transactions under the same pseudonym. $P \subseteq A \times E$ defines a profile, with E the stakeholder asserting the attributes. An attribute in a profile is considered *asserted* if an organization vouches for its correctness. Attributes that are considered *non-asserted* are supposed to originate from the user and their correctness is not vouched for by a trusted organization. For instance, a user's e-mail address is asserted if the service provider can verify its correctness. For that purpose, the user was obliged to click on a unique link he received in a special e-mail message. Formally, $(a_{e\text{-}mail}, o) \in P$ depicts an e-mail address that is asserted by organization o while $(a_{e\text{-}mail}, u) \in P$ represents the non-asserted variant. Profiles can be identifiable or pseudonymous:

- A profile P is **identifiable** if $\exists I \in \mathfrak{I} : \forall a \in I, \exists e \in E : (a, e) \in P$, with *identifiable set* $I \subseteq A$ and $\mathfrak{I}$ the set of all identifiable sets. These sets represent combinations of attributes that are required to sufficiently identify an individual person [15]. The term *sufficiently* is less restrictive than *uniquely*. For instance, someone's first name and surname can be considered to form an identifiable set, although, it is not unlikely to find two people with the same first name and surname.
- A profile P is **pseudonymous**, if $\exists N \in \mathfrak{N} : \forall a \in N, \exists e \in E : (a, e) \in P$, with *pseudonymous set* $N \subseteq A$ and $\mathfrak{N}$ the set of all pseudonymous sets. These sets contain attributes that form a unique combination. All data that is associated with it, can be linked. For instance, the set of attributes that represents the fingerprint of a browser [9] can be considered as a pseudonymous set.

A profile P where $\forall a \in I, \exists o \in O : (a, o) \in P$ is *asserted identifiable*. Similarly, a profile can be *asserted pseudonymous*. Multiple profiles can share a pseudonym or identifiable set. In this case all these profiles can be linked and merged together. A profile is *anonymous* if it is not identifiable, nor pseudonymous.

Authentication Technologies. Users mostly need to authenticate before they can consume a service. Authentication technologies T are classified into *claim-based technologies* and *network-based technologies*. In the former case, users authenticate using *credentials* that contain a set of attributes, such as anonymous credentials [13,3], certificates (e.g., X.509), or a username/password combination. Note that, compared to the other claim-based technologies, anonymous credentials enable the user to remain anonymous during authentication unless identifiable attributes are requested. To support accountability, authentication technologies allow to verify the authenticity of the credential's content (e.g., a digital signature guarantees that the content is authentic). $c \in C \subseteq 2^A \times O \times T$ defines a credential. The definition comprises a set of attributes $\{a, \dots\} \in 2^A$, the issuer $o \in O$, and the technology $t \in T$. Credentials are kept in the user's *credential wallet W*. Note that W can grow (e.g., after service consumption).

Network-based authentication technologies instead, are based on user profiles P that are managed by organizations. These can be accessed by the user and possibly external organizations. Usually, this requires authentication by one or more stakeholders E and possibly also the user's consent (e.g., she has to accept a profile access request). Hence, often these technologies are combined with claim-based authentication mechanisms. For instance, a user accesses a music stream service via his favorite social network account. Therefore, he needs to authenticate to the social network using his account name and password. Consequently, the music stream service gets read access to the public part of the user's social network profile and also the visible part of data of his friends in his social network. Furthermore, the application can update the music listening history in his profile that is shared among his social network friends.

Policies. Organizations enforce policies to access their services. These contain rules that define the conditions under which a service can be used. The policy rule-types that are considered in this ontology are *preconditions*, *post conditions*, *data storage*, and *data distribution*. Preconditions specify the attributes that must be revealed to an organization and the properties that must be fulfilled. It also includes whether data must be asserted. For instance, the user needs to reveal his name and date of birth to the MusicStreamInc and prove to be older than 18 before he is able to stream music to his computer. His age must be asserted and requires the consumer to use his identity card. Post conditions are causal rules that describe the output of services. For instance, the music stream subscription service issues a credential to a user. This credential provides access to the complete library of rock music. Data storage rules define the data that is stored in one or more profiles owned by the organization. Data distribution rules specify to which parties data is forwarded.

Each policy applies to a service $(s, o) \in \Sigma$ and is represented as $\delta_{(s,o)} \in \Delta_o$, with Δ_o the set of all policies of organization o. Possibly, alternative conditions for service access exist. For instance, during subscription the user can choose between two different social networks from which personal information is obtained. Different attributes are revealed to MusicStreamInc depending on the selection. Each policy is a disjunction of alternative sub-policies $\delta_{(s,o)} \equiv \delta_{(s,o)}^1 \vee \ldots \vee \delta_{(s,o)}^n$, with $n \in \mathbb{N}_0$.

Trust Relations. The data that organizations exchange depend on their trust relations. Users and organizations often have conflicting requirements. The former want minimal data disclosure. This view is based on the data minimization principle (i.e., to reveal the least amount of personal information). The more data a user reveals, the higher the risk to harm his privacy. The latter are profit driven and therefore they are often interested in building large profiles. This enables them to improve the level of personalization and make their applications more attractive to the user. Different trust relations are applied in this ontology:

- **Users Need to Trust Organizations.** We focus on *storage trust* relation $R_S \subseteq O$ and *distribution trust* relation $R_D \subseteq O$. Typically, organizations specify this behavior in a privacy policy that is publicly accessible to their users. Users have to accept it before they are granted access to the organization's services. Consequently, it is convenient to specify the user trust towards the organization's $o \in O$ policy set Δ_o. We assume that users do not trust other organizations by default ($o \notin R_S$ and $o \notin R_D$). This means that users really opt-in new organizations $o \in O$ in R_S and R_D. Expressing that a user trusts one of these policies assumes that only the specified data is affected by the services. We assume that untrusted organizations will store or forward all data that can be retrieved. For instance, an organization's policy can stipulate that the organization only obtains the user's name from his electronic identity card that uses the X.509 technology. If $o \in R_S$, then only the name is stored by the organization. Else, all the attributes of the user's certificate are retrieved and stored, although the policy might specify it differently.
- **Interactions Require also the Trust From Organizations Towards Users.** If data is not asserted, users can provide false information to organizations. For instance, they can enter an incorrect name and address at registration. To exclude inaccurate data, organizations can oblige users to provide asserted data. These trust relations are specified implicitly in the organization's access policy. The trust relation is defined as $R_\Sigma \subseteq A \times O \times \Sigma = \{(a, o_a, (s, o_s)) \mid (s, o_s) \in \Sigma$ and a precondition rule of $\delta_{(s,o_s)}$ specifies that o_a asserts attribute $a\}$.
- **Finally, Trust Is Often Required Between Interacting Organizations.** Two types of interactions are defined, namely (1) an organization can accept credentials issued by another organization or (2) an organization can forward data to another organization. Both are implicit in the organization's policies. The former are expressed in the access policy while the latter are expressed in the data distribution rules of the policy. The definition of this trust relation is $R_O \subseteq O \times O = \{(o_1, o_2) \mid \Delta_{o_1}$ specifies a rule where o_1 accepts credentials or profiles from o_2, or forwards data to $o_2\}$.

4   A Travel Reservation System

The approach is validated through the modeling of a travel reservation system. First, we give an overview of the system. Next, the system is modeled. Finally, we focus on valuable feedback that can be extracted from the model.

4.1   Scenario and Setup

A student wants to use an online travel reservation system to book a touristic trip. The travel agency offers triple-packs including the flight, hotel, and a theme park visit. Each item can also be booked separately. The latter requires users to access the web services of the airline company, the hotel chain and the theme park directly. Special reductions are offered at the theme park's website to students and loyal hotel guests. The airline company and the theme park belong to the same holding. They collaborate to increase efficiency. The student believes that the travel reservation system stores (and can forward) all data released to that organization (even when specified differently in its privacy policy). Also, he has only limited trust in the hotel chain (i.e., he assumes that it stores all collected data). Before he books his trip, he wants to have an insight on the profiles that can be built by the organizations when using a specific booking strategy (*fb1*). Furthermore, he wants to know the most privacy-preserving strategy to book the theme park tickets (*fb2*). Finally, he checks whether he remains unidentifiable by the theme park when taking a student offer (*fb3*).

Services and organizations. The user $u \in U$ and the organizations $O = \{o_{travel}, o_{air}, o_{hotel}, o_{park}, o_{gov}, o_{univ}\}$ are the stakeholders in the system. Note that the holding only supports collaboration between the airline company and the amusement park. The government o_{gov} issues governmental electronic identity cards and electronic driving licenses, while an educational institution o_{univ} issues the electronic student card. Table 1 gives an overview of the authentication technologies together with the attributes that are included. All credentials are X.509 certificates except the hotel's rewards membership credential c_{rew}, which is a

Table 1. Travel reservation system authentication

Credential	x	c_x		Issuer
		$a \in A$		
eID	eID	name, address, citizenship, DoB, profession, SSN		o_{gov}
Driving license	driv	name, address, citizenship, DoB		o_{gov}
Student card	stud	name, address, DoB, institute, study		o_{univ}
Hotel rewards	rew	reward id		o_{hotel}

Profile	x	P_x		Owner
		$A \times E$		
Hotel rewards	rew	(reward id, o_{hotel}), (name, u), (e-mail, o_{hotel}), (DoB, u), (rewards status, o_{hotel})		o_{hotel}

username/password combination. The latter gives access to a membership account P_{rew}. The services Σ are split in two groups Σ' and Σ'', with $\Sigma = \Sigma' \cup \Sigma''$. The former are directly accessible by consumers (i.e., $\Sigma' = \{(book, o_{travel}),$ $(book, o_{air}), (book, o_{hotel}), (book, o_{park}), (reduction, o_{park})\}$) while the latter can only be accessed by the travel agency (i.e., $\Sigma'' = \{(book\_ext, o_{air}), (book\_ext, o_{hotel}), (book\_ext, o_{park})\}$).

Policies and trust relations. The travel agency $(book, o_{travel})$ defines that a consumer must at least prove his name and address before it can book a triple-pack

Table 2. Travel reservation system policies model of the user accessible services

(s, o)	i	$\delta^i_{(s,o)}$
$(book, o_{travel})$	1,2	**reveal**(name, address, citizenship, DoB) **from** c_{eID}1 **or** c_{driv}2 **reveal**(e-mail, flight destination, date of travel, diet, hotel location, room type, date of arrival, date of departure) **from** u **reveal**(reward id) **from** c_{rew} **forward**(name, address, citizenship, DoB, e-mail, flight destination, date of travel, diet) **to** o_{air} **scope** $(book\_ext, o_{air})$ **forward**(name, address, citizenship, DoB, e-mail, hotel location, room type, reward id, date of arrival, date of departure) **to** o_{hotel} **scope** $(book\_ext, o_{hotel})$ **forward**(e-mail, date of visit) **to** o_{park} **scope** $(book\_ext, o_{park})$
$(book, o_{hotel})$	1	**reveal**(name, address, citizenship, DoB) **from** c_{eID} **reveal**(reward id) **from** c_{rew} **reveal**(e-mail, hotel location, room type, date of arrival, date of departure) **from** u **store**(name, address, citizenship, DoB, e-mail, hotel location, room type, reward id, date of arrival, date of departure) **output**(c_{hotel}) **by** o_{hotel}
$(book, o_{air})$	1	**reveal**(name, address, citizenship, DoB) **from** c_{eID} **reveal**(e-mail, flight destination, date of travel, diet) **from** u **store**(name, address, citizenship, DoB, e-mail, flight destination, date of travel) **output**(c_{air}) **by** o_{air}
$(book, o_{park})$	1	**reveal**(e-mail, date of visit) **from** u **store**(e-mail, date of visit) **output**(c_{park}) **by** o_{park}
$(reduction, o_{park})$	1	**reveal**(e-mail, date of visit) **from** u **reveal**(name, address, DoB) **from** c_{stud} **store**(name, address, DoB, e-mail, date of visit) **output**(c_{park}) **by** o_{park}
	2	**reveal**(e-mail, date of visit) **from** u **reveal**(reward id) **from** c_{rew} **reveal**(rewards status) **from** P_{rew} **store**(reward id, e-mail, date of visit) **forward**(reward id) **to** o_{hotel} **scope** $(get\ rewards\ status, o_{hotel})$ **output**(c_{park}) **by** o_{park}

Table 3. Travel reservation system policies model of the services that are accessible for external organizations

(s, o)	i	$\delta^i_{(s,o)}$
$(book\_ext, o_{hotel})$	1	**reveal**(name, address, citizenship, DoB, rewards id, e-mail, hotel location, room type, date of arrival, date of departure) **from** o_{travel} **store**(name, address, citizenship, DoB, e-mail, hotel location, room type, reward id, date of arrival, date of departure) **output**(c_{hotel}) **by** o_{hotel}
$(book\_ext, o_{air})$	1	**reveal**(name, address, citizenship, DoB, e-mail, flight destination, date of travel, diet) **from** o_{travel} **store**(name, address, citizenship, DoB, e-mail, flight destination, date of travel) **output**(c_{air}) **by** o_{air}
$(book\_ext, o_{park})$	1	**reveal**(e-mail, date of visit) **from** o_{travel} **store**(e-mail, date of visit) **output**(c_{park}) **by** o_{park}

using the $(book\_ext, o)$ service. Those attributes must be asserted by the government using an electronic identity card or driving license. Other attributes, such as his e-mail address and flight destination do not need to be asserted. Only personal data that is strictly necessary to obtain a plane voucher c_{plane}, a hotel voucher c_{hotel}, and a theme park voucher c_{park} are forwarded to the respective organizations . The airline company and hotel chain require at least the user's name and his address when tickets are booked directly using $(book, o_{air})$ and $(book, o_{hotel})$. These must be asserted by the government via the identity card. Other attributes, such as the user's e-mail address, hotel location, and flight destination are non-asserted. All attributes are stored by both organizations. Optionally, tourists can also define diet preferences. Both organizations issue a voucher to the consumer after a successful reservation. A theme park visit $(book, o_{park})$ only requires the user's e-mail address and the date of visit (non-asserted). The visitor gets an entry voucher. All his attributes are stored. Tourists get a reduction when they can prove to be a student or to be silver members in the hotel's rewards program. When the student uses his student card to fulfill the preconditions, the theme park obtains the name, address, and date of birth (DoB) that are asserted by the student's school. The consumer can also opt to prove to be a silver member. If so, he is redirected to the hotel chain where he needs to login in with his password. After a successful login, the user's *rewards status* is forwarded to the theme park. Attributes that are released to get a reduction are stored by the theme park. Table 2 and 3 give a formal overview of the policies that are applied in the travel reservation system.

Multiple trust assumptions can be applied. $R_S = \{o_{air}, o_{park}\}$ means that the user trusts the storage policies of the airline company and theme park. $R_D = \{o_{air}, o_{hotel}, o_{park}\}$ means that the user trusts the distribution policies of the organizations included in R_D. Note that it is quite trivial to modify these sets.

4.2 Feedback

To demonstrate the expressive power of our approach, a set of queries are performed on the formal model, namely *fb1, fb2 and fb3*.

Table 4. Travel reservation system user profiles based on the user trust in organizations

Profile	Owner	Attributes	Asserted
$P^{eID}_{(book,o_{travel})}$	o_{travel}	name, address, citizenship, DoB, profession, SSN	o_{gov}
		reward id	o_{hotel}
		e-mail, flight destination, date of travel, diet, hotel location, room type, date of arrival, date of departure	u
$P^{driv}_{(book,o_{travel})}$	o_{travel}	name, address, citizenship, DoB	o_{gov}
		reward id	o_{hotel}
		e-mail, flight destination, date of travel, diet, hotel location, room type, date of arrival, date of departure	u
$P_{(book,o_{hotel})}$	o_{hotel}	name, address, citizenship, DoB, profession, SSN	o_{gov}
		reward id	o_{hotel}
		e-mail, hotel location, room type, date of arrival, date of departure	u
$P_{(book,o_{air})}$	o_{air}	name, address, citizenship, DoB	o_{gov}
		e-mail, flight destination, date of travel, diet	u
$P_{(book,o_{park})}$	o_{park}	e-mail, date of visit	u
$P^{stud}_{(reduction,o_{park})}$	o_{park}	name, address, DoB	o_{univ}
		e-mail, date of visit	u
$P^{rewards}_{(reduction,o_{park})}$	o_{park}	reward id, rewards status	o_{hotel}
		e-mail, date of visit	u
$P^{eID}_{(book\_ext,o_{hotel})}$	o_{hotel}	name, address, citizenship, DoB, profession, SSN	o_{gov}
		reward id	o_{hotel}
		e-mail, flight destination, date of travel, diet, hotel location, room type, date of arrival, date of departure	u
$P^{driv}_{(book\_ext,o_{hotel})}$	o_{hotel}	name, address, citizenship, DoB	o_{gov}
		reward id	o_{hotel}
		e-mail, flight destination, date of travel, diet, hotel location, room type, date of arrival, date of departure	u
$P_{(book\_ext,o_{air})}$	o_{air}	name, address, citizenship, DoB	o_{gov}
		e-mail, flight destination, date of travel, diet	u
$P_{(book\_ext,o_{park})}$	o_{park}	e-mail, date of visit	u

fb1: User profiles. The end-user and/or designer can get an overview of the information spreading in the system. Details are provided in Table 4. Each service consumption (s, o) leads to a profile $\delta^{(i)}_{(s,o)}$ containing the attributes gathered by the service. The specific booking strategy has an impact on the type and amount

of attributes that are part of the profile. $P^{eID}_{(book,o_{travel})}$ and $P^{driv}_{(book,o_{travel})}$ show the attributes in the profile when the identity card or the driving license are used, respectively. The student only has very limited trust in the travel agency's storage policy and therefore the profile contains all attributes that are released to that organization. Among others, a social security number (SSN) and profession are added to $P^{eID}_{(book,o_{travel})}$ as these are the attributes of the X.509 certificate in the eID card.

The travel agency needs to forward attributes to the airline company, hotel chain, and theme park. In return, vouchers are issued. The student assumes that the travel agency forwards all data that is gathered. The profiles $P_{(book\_ext,o_{air})}$ and $P_{(book\_ext,o_{park})}$ only contain the attributes that are specified in the respective storage policies, as the tourist trusts their storage policies. The profile that is kept by the hotel chain contains all data that is released by the user. Using the driver's license for authentication leaks less information than using the eID card. More specifically, the user's SSN and profession are stored in the profile when the eID is used. The profiles $P_{(book,o_{air})}$, $P_{(book,o_{hotel})}$, and $P_{(book,o_{park})}$ list the set of attributes that are collected if the travel agency does not mediate in the bookings. All profiles comply with the specified storage policies, except the one owned by the hotel. The hotel keeps the user's SSN and profession. Note further, that the strategy that is selected to get a reduction at the theme park also has an impact on the attributes in the profile.

fb2: Impact of reduction on privacy. Selecting a specific booking strategy can have an impact on the user's privacy. If the travel agency is used, authenticating with the driver's license better protects the privacy than using an eID card. However, the travel agency still collects a lot of valuable personal data. No data is released to the travel agency in case of direct bookings, but this compels the consumer to use his identity card for the hotel and airplane booking. Hence, the user's SSN is revealed to both organizations. The latter is a unique identifier to which a lot of information can be linked. If a student wants a reduction, he must book directly. At first glance, using the rewards membership seems the most privacy-friendly option as only the user's e-mail address and reward id are revealed. In contrast, the user's e-mail address, name and address are revealed if the student card is used. However, the theme park and the airline company belong to the same holding and exchange data. The profiles from the airline company and theme park can be linked by means of the user's e-mail address. Thus, the theme park also knows the user's name and address. Hence, using the student card is slightly more privacy-friendly.

fb3: Reductions are identifiable. Verifying if the student can remain unidentifiable towards the theme park when booking a voucher at reduced price is done by a query based on identifiable sets. The identifiable sets in our system are $\mathfrak{I} = \{I_1, I_2\}$, with $I_1 = \{name\}$ and $I_2 = \{address\}$ where *name* comprises the user's name and surname. Note that the user or designer can define what is *identifiable*. Note that the union $I_3 = I_1 \cup I_2$ is also an identifiable set. A user is unidentifiable to an organization if the following two conditions are met.

First, the organization owns no identifiable profile. Second, the organization contains no pseudonymous set that is shared with an identifiable profile of a collaborating organization. According to these conditions, the first condition is met only when the student uses his reward id. However, the second condition is not fulfilled. To explain this, the pseudonymous sets $N_1 = \{e\text{-}mail\}$ is required. $N_1 \subseteq P_{(book,o_{air})} \cap P^{rewards}_{(reduction,o_{park})}$ and $I_3 \subset P_{(book,o_{air})}$ (profile is identifiable). Hence, booking at reduced price is identifiable.

5   Evaluation

The presented approach is holistic in the sense that a wide range of authentication technologies, multiple service policies and trust assumptions, and advanced interactions are supported. The formal approach facilitates the concrete realization using existing first-order logic tools. These enable the automatic extraction of feedback to inspect the user's privacy. IDP was selected for the realization [5] and was applied to the travel agency scenario. To evaluate the privacy, different types of automated feedback are used. Therefore, profiles of each organization – that are based on the service policies and the user's trust relations – are compiled. These profiles are examined for linkage with pseudonyms and identities. The results are part of the feedback [5]. Because of the amount of data, the output lacks readability. The view on privacy in this approach is from the user's viewpoint because it is based on the user's trust assumptions. Other aspects in the model that have an impact on information spreading are collaboration between organizations and the used authentication technologies. Only the information spreading is considered by this approach and not the goal of the collection of data. Therefore, other complementary approaches [18] can be used. The model provides a static view on the privacy. Consequently, sequences of events over time that influence the user's privacy cannot be expressed. However, this significantly reduces the search space for finding results on different types of feedback. Furthermore, stronger conclusions are possible when omitting time because these conclusions are not restricted to a finite time-domain.

Systems are modeled from a single-user viewpoint. This enables to abstract the actual values of attributes (included in credentials and profiles). For instance, *John lives in New York* is a real-world proposition. Our modeling approach abstracts the actual values of the attributes *name* and *city*. However, a mapping table that contains tuples *(attribute_type, attribute_value)* can easily be added to instantiate attributes in case it is not feasible to abstract away the attribute value. Abstracting attribute values may also complicate the modeling of graph structures. For instance, a friend can be part of a user's social network. A predicate $MemberOf(Friendlist, Friend)$ can be used to express this property.

Real-world service policies describe how organizations deal with personal data. The service policy specifications in our approach consider the same information specified in formal rules. Although, such policies are often vague and insufficient to extract the exact behavior of organizations, there exist policy specification languages that facilitate service providers to express their policies in detail. For instance, CARL [2] can be used to express access control requirements. Here, the model's service policy rules can be extracted automatically.

The approach supports different types of modifications in the system. Changes in service policies can be handled automatically in case they are specified in languages such as CARL. Otherwise, modifications in the service policies require manual interventions. Other types of changes, such as organization collaborations and newly added organizations are also added manually to the model. For instance, adding a new organization o to the system involves adding the new organization to the set of stakeholders E and the set of organizations O. New services are added to the sets Σ and S and new service policies of organization o are added to Δ_o.

Future research will focus on the integration in design tools, for instance for the compliance verification of service policies with corporate level and governmental policies. This approach can also be applied to end-user applications. For instance, a browser plug-in to inspect if the data that is revealed complies with the user defined policy, such as his SSN that must be kept hidden. This involves an on-line model of the system that can be consulted from the browser plug-in. Furthermore, improving the readability of the feedback output is future work.

6 Conclusions

This paper presented a formal approach to model advanced electronic services in which multiple organizations are involved. Once a system is modeled, the designer and/or end-user can inspect privacy properties based on varying trust assumptions. By the design of a travel reservation system, we show the feasibility of the approach and demonstrate how different types of feedback are extracted from the formal model.

References

1. Barth, A., Datta, A., Mitchell, J.C., Nissenbaum, H.: Privacy and contextual integrity: Framework and applications. In: Proceedings of the 2006 IEEE Symposium on Security and Privacy, SP 2006, pp. 184–198. IEEE Computer Society, Washington, DC (2006)
2. Camenisch, J., Mödersheim, S., Neven, G., Preiss, F.-S., Sommer, D.: A card requirements language enabling privacy-preserving access control. In: Proceedings of the 15th ACM Symposium on Access Control Models and Technologies, SACMAT 2010, pp. 119–128. ACM, New York (2010)
3. Camenisch, J., Van Herreweghen, E.: Design and implementation of the idemix anonymous credential system. In: Proceedings of the 9th ACM Conference on Computer and Communications Security, CCS 2002, pp. 21–30. ACM, New York (2002)
4. Microsoft Corporation. Privacy guidelines for developing software products and services, version 3.1 (September 2008),
http://www.microsoft.com/en-us/download/details.aspx?id=16048
5. Decroix, K.: Inspect privacy and trust (2012),
http://code.google.com/p/inspect-privacy-and-trust/

6. Denecker, M.: Extending Classical Logic with Inductive Definitions. In: Palamidessi, C., Moniz Pereira, L., Lloyd, J.W., Dahl, V., Furbach, U., Kerber, M., Lau, K.-K., Sagiv, Y., Stuckey, P.J. (eds.) CL 2000. LNCS (LNAI), vol. 1861, pp. 703–717. Springer, Heidelberg (2000)

7. Deng, M., Wuyts, K., Scandariato, R., Preneel, B., Joosen, W.: A privacy threat analysis framework: supporting the elicitation and fulfillment of privacy requirements. Requirements Engineering 16, 3–32 (2011)

8. Díaz, C., Seys, S., Claessens, J., Preneel, B.: Towards Measuring Anonymity. In: Dingledine, R., Syverson, P. (eds.) PET 2002. LNCS, vol. 2482, pp. 54–68. Springer, Heidelberg (2003)

9. Eckersley, P.: How Unique Is Your Web Browser? In: Atallah, M.J., Hopper, N. (eds.) PETS 2010. LNCS, vol. 6205, pp. 1–18. Springer, Heidelberg (2010)

10. Langheinrich, M.: Privacy by Design - Principles of Privacy-Aware Ubiquitous Systems. In: Abowd, G.D., Brumitt, B., Shafer, S. (eds.) UbiComp 2001. LNCS, vol. 2201, pp. 273–291. Springer, Heidelberg (2001)

11. Mariën, M., Wittocx, J., Denecker, M.: The IDP framework for declarative problem solving. In: Search and Logic: Answer Set Programming and SAT, pp. 19–34 (2006)

12. Naessens, V., De Decker, B.: A Methodology for Designing Controlled Anonymous Applications. In: Fischer-Hübner, S., Rannenberg, K., Yngström, L., Lindskog, S. (eds.) SEC 2006. IFIP, vol. 201, pp. 111–122. Springer, Boston (2006)

13. Paquin, C.: U-prove technology overview v1.1 draft revision 1. Microsoft Corporation (February 2011)

14. Pearson, S.: Privacy Management in Global Organisations. In: De Decker, B., Chadwick, D.W. (eds.) CMS 2012. LNCS, vol. 7394, pp. 217–237. Springer, Heidelberg (2012)

15. Pfitzmann, A., Hansen, M.: A terminology for talking about privacy by data minimization: Anonymity, unlinkability, undetectability, unobservability, pseudonymity, and identity management, v0.34 (August 2010)

16. Serjantov, A., Danezis, G.: Towards an Information Theoretic Metric for Anonymity. In: Dingledine, R., Syverson, P.F. (eds.) PET 2002. LNCS, vol. 2482, pp. 41–53. Springer, Heidelberg (2003)

17. Sindre, G., Opdahl, A.L.: Eliciting security requirements with misuse cases. Requirements Engineering 10, 34–44 (2005)

18. Tschantz, M.C., Datta, A., Wing, J.M.: Formalizing and enforcing purpose restrictions in privacy policies. In: IEEE Symposium on Security and Privacy, pp. 176–190. IEEE Computer Society (2012)

19. Veeningen, M., de Weger, B., Zannone, N.: Formal Privacy Analysis of Communication Protocols for Identity Management. In: Jajodia, S., Mazumdar, C. (eds.) ICISS 2011. LNCS, vol. 7093, pp. 235–249. Springer, Heidelberg (2011)

20. Wittocx, J., Mariën, M., Denecker, M.: The IDP system: a model expansion system for an extension of classical logic. In: LaSh, pp. 153–165 (2008)

Idea: Writing Secure C Programs with SecProve

Myla M. Archer, Elizabeth I. Leonard, and Constance L. Heitmeyer*

Naval Research Laboratory (Code 5546)
Washington, DC 20375

Abstract. This paper describes SecProve, a prototype tool we are developing for checking application-specific security properties of C code, together with our vision of how such a tool can be used by a programmer to maintain security of code during its development.

Keywords: security, verification, security properties, support for assurance, security best practices, code security, application-specific security.

1 Introduction

Many serious vulnerabilities in systems arise from security violations in software. Unfortunately, software testing, the most common approach to detecting security violations, provides by itself little confidence that a program is secure. Although formal verification could significantly increase confidence in the security of software, it is viewed as too technically difficult, costly, and time consuming. Hence, demonstrating that a program is secure remains a challenging problem.

Recently, a number of powerful commercial tools, called static analyzers, have been introduced to address the problem of software security. Based on research in static analysis and similar techniques (see, e.g., [11], [10]), static analyzers, such as Coverity [1] and CodeSonar [8], detect *application-independent* errors (often called "code vulnerabilities") which do not depend on the application. Examples are null pointer dereferences, format string problems, integer range errors, and buffer overflows. Static analyzers have been applied to a large number of C, Java, C++, and C# programs and are estimated to have detected tens of thousands of bugs, most of which traditional software testing would not have found. Static analyzers have been highly successful because they are automatic and easy to use—applying the tools requires neither special skills nor special training.

To date, researchers and commercial tool vendors have largely ignored a second important class of security errors—*application-specific errors*—i.e., violations of security properties specific to the application. Examples include illegal data flows and failure of a program to sanitize memory areas after processing sensitive data in those areas. Gary McGraw, a leading computer security authority, estimates that, of the large number of security errors in current programs, approximately 50% belong to this second class [16]. However, detecting application-specific errors can be extremely difficult. Unlike application-independent errors, which static analysis tools can detect automatically and without user guidance, before a tool can detect application-specific errors, the developer must first define

---

* This research is supported by the Office of Naval Research.

J. Jürjens, B. Livshits, and R. Scandariato (Eds.): ESSoS 2013, LNCS 7781, pp. 171–180, 2013.
© Springer-Verlag Berlin Heidelberg 2013

the security properties of interest. Specifying these properties can be difficult, especially if the developer must express them in an unfamiliar language or logic.

This paper describes SecProve, a process and prototype tool, whose goal is to detect application-specific security violations in programs automatically. Unlike our earlier research [14,15], which used a model-based approach to verify the security of software, SecProve does not rely on a formal security model. Instead, it automatically checks a program for desired security properties and notifies the developer when a security violation is detected. An important feature of our approach is that SecProve checks for security violations as the developer is writing the program. Another important feature is that SecProve facilitates the specification of security properties by providing the developer with a set of templates. To make analysis by SecProve feasible, we assume that the C program to be analyzed is developed following best coding practices, e.g., no pointer arithmetic, no GOTOs, and no statements used as expressions. This paper reviews major concepts on which SecProve is based, including two—forward propagation of assertions and incremental compilation—supporting analysis of code during development, describes the SecProve process and tool support for developing secure code, presents an example to illustrate how the tool works, and concludes by describing our progress and future plans.

2   Background

Application-Specific Security Properties. Most application-specific security properties of code fall into well-known classes, such as sanitization, data flow, data influence, data integrity, data separation, access control, and non-bypassibility. The security properties proved of the certified software application in [14,15] include, for example, sanitization and data separation.

Unlike application-independent security properties, application-specific properties must be defined by the developer. Doing so requires detailed knowledge of the code design. For example, to define a sanitization property, the developer must identify the variables or memory areas to be sanitized, places in the code requiring their sanitization as a precondition, and the name of the procedure that implements sanitization. As noted above, to specify a desired security property in SecProve, the developer fills in a template. As an example, a template for specifying a sanitization property might provide slots for the property's ID; the data type of the item(s) to be sanitized; a sanitization predicate; a sanitization routine plus areas it sanitizes; and names of procedures that have sanitization of certain memory areas as a precondition, together with the relevant set of memory areas. Any class of security properties may further divide into subclasses, each with its own template. For data flow, for example, two subclasses are *forbidden* flows between two variables, and flows between two variables *restricted* to flow through a third variable.

LEMA. LEMA [3,4] is an abstract language for representing an imperative program in an intermediate form useful for reasoning about the program. It is basically a strongly typed language of `while` programs with support for procedures and annotation with assertions. The LEMA translator converts LEMA

procedures and assertions into state machines and candidate invariants defined
in the language of the PVS theorem prover [18]. Type checking of LEMA programs and assertions is done in PVS; checking of candidate invariants is done
using either the PVS theorem prover or any SMT solver supported in PVS.
As illustrated in Section 4, LEMA notation resembles the notation of PASCAL
rather than C notation. For example, equality in LEMA is represented by "="
rather than "==", and assignment by ":=" rather than "=".

Assertion Propagation. Reasoning about imperative programs is generally
done through the propagation of assertions in the code. Either weakest preconditions (resulting in backward propagation) or strongest postconditions (resulting
in forward propagation) are computed, with loops handled using loop invariants.
To avoid existential quantifiers in forward propagation, logical variables may be
introduced. Though loop invariants can sometimes be found automatically, in
general, the problem of finding loop invariants sufficient for proving desired program properties is undecidable. Our initial focus is straight line code, including
code with unwindable loops, because proving security properties about simple
straight line code has proven extremely useful (see, e.g., [14,15]). Generating assertions automatically for such code is possible using either backward or forward
propagation.

Backward propagation, typically used to verify that code satisfies a given
postcondition provided it satisfies a given precondition, is the most frequently
used technique for verifying the functional correctness of programs. However,
forward propagation, used for example by [17,12], has advantages for verifying
security properties. Establishing these properties can require the verification of
a set of desired code assertions derived from a property specification, rather
than verification of a limited set of postconditions which can be backward propagated. Because assertions generated by forward propagation are known to be
valid, they can be used to check desired code assertions in a procedure even before the procedure code is complete. Given that our goal is to provide feedback
to the programmer about security violations as soon as possible during code
development, our approach is to use forward propagation to the extent feasible.

Incremental Parsing. To better support analysis of code as it is developed, we
plan to support code changes while minimizing any re-analysis of the unchanged
portion of the code. To do this, we will build on existing results in incremental
parsing, as described, for example, in [13,7].

3   Approach to Tool Support

As described above, our goal is to create tool support for specifying desired
application-specific security properties of software and for checking the code
during development for property violations. The central idea is to base the code
analysis on *code assertions*, i.e., assertions annotating the code. SecProve will
support analysis of source code in the C language (and potentially other imperative languages). Often source code, especially C code, is not formatted to fully
support annotation: for example, there may be multiple commands on a line,

and some constructs may represent both commands and expressions. For this reason, SecProve uses an intermediate LEMA representation of the source code during code analysis.

LEMA code is formatted so that a new line begins at every point in the code where a change in the program state may occur, e.g., after an assignment, a procedure call, or a change in control location after a test. In analogy with an `assert` statement in C code, any assertion associated with a line of LEMA code refers to the program state when that line is reached—and before it is executed. In SecProve, code assertions associated with lines in the intermediate LEMA representation of C code derive from one of three sources: 1) assertions and contracts which the developer has included with the C code, 2) security properties specified by the developer, and 3) inference from the (LEMA) code itself. Because the number of assertions associated with a given line of LEMA code may be very large, SecProve does not interleave the assertions with the LEMA code, but keeps them in a separate database. Nor does SecProve ever modify or annotate the developer's C code.

Assertions that a code developer includes in source code typically express facts that the developer desires to be true at certain points in the code during execution. Alternatively, the developer may use assertions to specify desired contracts for functions or procedures. A compiler can use such assertions to specify checks to be done during execution, and halt execution when they fail. In contrast, the assertions used for code analysis in SecProve fall into two categories:

- *desired* assertions, including assertions which, if proved to hold, guarantee specific security properties; and
- *valid* assertions derived from the code or from verification of some desired assertion.

In SecProve, tool support will:

- Maintain a database of information about the code and its desired security properties, for use in analyzing the code with respect to the properties;
- Provide the software developer with templates for specifying common classes of application-specific security properties;
- Compute code assertions of two classes, *Class 1* and *Class 2*—where Class 1 assertions either do not require checking or have already been checked, and Class 2 assertions still require checking—and enter the assertions in the database to support code analysis;
- Provide theorem proving support for checking Class 2 assertions; and
- Provide feedback to the developer about the state of the code development and analysis, including traceability information from the intermediate representation to the original source code to connect any detected property violation to its point of origin in the source code.

Information about the code to be maintained in the database includes the abstract syntax tree and control flow graph of the LEMA code; traceability information from LEMA code to source code; code assertions represented either directly or, when the potential number of assertions is very large (for example, in the case of assertions about data flow or data influence), in some more compact tabular form; dependencies among assertions; any needed verification conditions; and so on. Information about desired security properties of the code will be represented in the database as instantiated property templates.

For code analysis, the following assertions will be automatically generated:

- For a given property or set of properties, desired code assertions which, if valid, ensure that the property is not violated;
- Assertions which can be deduced as valid from the code itself; and
- A desired precondition and a postcondition, valid upon verification of the desired precondition, for every procedure call in the code, and derived by instantiating the precondition and postcondition in the procedure's contract with the actual parameters in the procedure call.

Class 2 assertions guaranteeing a particular security property can be generated as soon as that property has been specified and entered in the database. Class 1 assertions obtained by forward propagation through the LEMA code can be entered into the database during translation of source code to LEMA code, as can those Class 2 assertions that are preconditions of procedure calls.

The database will be used to provide both 1) the information needed by a theorem prover or SMT solver to check the validity of the Class 2 assertions and, if possible, convert them to Class 1 and 2) the information needed to display the current state of the code development and analysis to the developer, and to accept input (property specifications and new code) to the analysis tools through a GUI. Figure 1 illustrates our concept of the developer's view of an Eclipse-like GUI. Code analysis—which entails modifying the database and validity checking of assertions—will occur only upon user request through the GUI, because it can be expensive computationally. To handle the potential explosion in the number of assertions available for theorem proving, the developer can constrain the analysis—e.g., by focusing on certain predicates and variables.

Because a programmer may make changes at arbitrary points in the C source code during its development, the SecProve database (see [9]) is designed to facilitate such code changes. The database associates an unchanging unique line *ID* (distinct from the line *number*) with each line of LEMA code when it is first generated from the C code. Each piece of information associated with a particular line of code—such as C source line number, LEMA code line number, variable scope information, or Class 1 or 2 assertion—is associated in the database with the unique line identifier of that line of code. Thus, for example, assertions associated with a line of LEMA code that is replaced or deleted will become inaccessible. A change in the C code will produce a corresponding change in the LEMA representation, and may require re-analysis of the code. To minimize the re-analysis effort, the stored procedures which manage the database will maintain as much of the information associated with unaltered lines of LEMA code as possible.

4   Example: Checking a Sanitization Property

This section illustrates how a developer would use SecProve to check a sanitization property and a developer-supplied contract for a simple C program.

Figure 1 illustrates the current design of the SecProve GUI. The middle upper window of the GUI displays the current C code, which is not yet complete: the function get_input is only a stub, and no code yet exists for the functions processing and write_random. Because it is incomplete, this program will not compile. SecProve, by contrast, can analyze this (partial) program.

SecProve provides two methods for developers to specify desired properties: 1) contracts and assertions placed directly in the C code (described later in this section), and 2) templates for specifying application-specific security properties. Displayed in the lower right corner of Figure 1 is a sanitization property template. In the example, the security property of interest is that sanitized(partition) holds whenever the procedure process_data begins execution. To define this property using the template, the developer provides: 1) a name for the property (sanitize_partition); 2) the name of a routine that performs sanitization (cleanup), together with the set of global variables or formal parameters it sanitizes ({partition}); 3) the data type subject to sanitization (int _[10], i.e., integer arrays of size 10); 4) the name of a predicate that indicates whether a data area of the given data type is sanitized (sanitized); and 5) a list of those procedures requiring as a precondition that certain data areas are sanitized (process_data), together with the set of global variables or formal parameters assumed sanitized ({partition}). The template can be instantiated as many times as necessary to cover every data area for which sanitization is a concern, e.g., because it may hold sensitive data.

As noted in Section 3, SecProve supports developer-supplied annotations in the form of function contracts and assertions at individual lines of code. While C assert statements can be used to annotate individual lines of code, SecProve, like Boogie [5], provides two further annotation constructs, requires and ensures, to respectively capture contract preconditions and postconditions. SecProve expects a contract for a C function, if provided by the developer, to appear immediately after the opening curly brace for the function body. Placing the contract at the beginning of the function allows the contract to be used in analysis as early as possible, including situations in which the code for the function is incomplete or missing entirely. A function with only a function declaration and a contract is called a *stub*. Contracts associated with stubs can be used in the analysis of other routines, with proof that the stub satisfies its contract postponed until the code for the stub is developed. The example C code in Figure 1 includes a developer-supplied contract ensures(sanitized(partition)) for process_data indicating that sanitized(partition) is desired to hold at the end of process_data. SecProve automatically generates a header file, security_predicates.h, containing trivial definitions (returning true) for requires and ensures. From the property templates, SecProve automatically generates trivial definitions in security_predicates.h for any predicates (e.g., sanitized) declared in property specifications, so that the developer can use these predicates in annotations. This header file allows the C code to compile. To allow ensures annotations to refer to both the beginning and ending values of a variable passed to a routine as an actual parameter, SecProve automatically adds to the header file the declaration of a variable x_save for each global variable or formal parameter x in the program. This variable refers to x's value when a

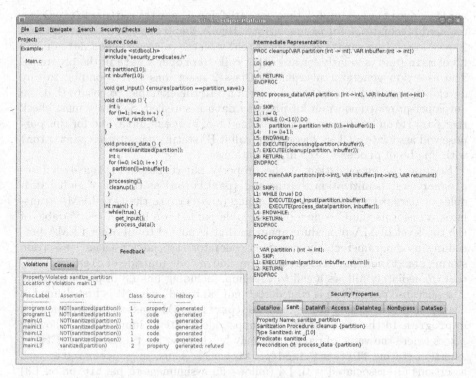

Fig. 1. Conceptual SecProve GUI showing sanitization example

routine is first entered. Thus, in the C program in Figure 1, `partition_save` can be (and is) used in the `get_input` stub `ensures(partition == partition_save)` to indicate that `get_input` does not change the value of `partition`. Because this is a stub, no proof is required until the stub is expanded to include code.

The first step of SecProve's verification process is to translate the C code into an equivalent program in LEMA. The GUI displays the LEMA representation of the C code in the window in the upper right corner. To display all of the sanitization-relevant code in Figure 1 at the same time, the LEMA representation of `get_input` has been removed and portions of the LEMA representation of other procedures have been replaced by ellipses. During the generation of the LEMA intermediate representation, information is added to the database (e.g., about the developer-supplied assertions and contracts). The security properties specified using the templates are also added to the database. Each property is assigned a unique property ID used as a key in the database to associate the property with assertions that will be generated from the property.

Next, for each specified property, the assertion generator processes the property specification and associates, with both the property and appropriate code locations, Class 2 assertions which, if they all hold, are sufficient to guarantee the property. All assertions are stored in the database by associating (but not interleaving) them with the LEMA code. The specification of property sanitize_partition implies that, for satisfying it, sanitized(partition) is a required precondition for process_data. Hence, SecProve will associate a Class 2 assertion stating sanitized(partition) with the call to process_data on line L3 of main (and, similarly for any other calls to process_data in the program). The assertion generator also enters Class 2 assertions corresponding to any developer-provided contracts and assertions. In the example, to ensure that the developer-provided contract for process_data is satisfied, SecProve must check that partition has been sanitized at line L8 of process_data, and for this purpose will associate a Class 2 assertion (call it B) stating sanitized(partition) with line L8 of process_data in the database.

Because the specification of the property sanitize_partition designates cleanup as its sanitization routine and {partition} as the set of global variables or parameters sanitized by cleanup (and because the C-to-LEMA translator transforms the C code global variable partition into a local variable of the top level LEMA procedure program that is passed to all other LEMA procedures as a parameter), the assertion generator associates a Class 1 assertion stating sanitized(partition) with any line of code (outside of cleanup itself) that immediately follows a call to cleanup, including line L8 in process_data. Also, because partition is considered "not sanitized" when it is first defined, a Class 1 assertion stating NOT(sanitized(partition)) is associated with line L0 in program. In the code outside cleanup, code locations immediately following places where the value of partition may be changed are tagged with a Class 1 assertion stating NOT(sanitized(partition)). Thus, in process_data, such assertions are associated with L4 (follows an assignment to partition on L3) and with L7 (follows a call to processing on L6). Note that because the behavior of processing with respect to partition is unknown, it must be assumed that partition is not sanitized after processing executes.

Let A be the Class 1 assertion associated with line L8 in process_data. Since A is equivalent to the Class 2 assertion B at L8 of process_data, B can be converted to Class 1 and marked as proved. This proves that the developer-supplied contract for process_data holds. The dependency of B's Class 1 status on A is recorded in the database so that if updates to the code eliminate A, B will revert to Class 2 in the database.

By forward propagation of the Class 1 assertion stating NOT(sanitized(partition)) associated with line L0 in program, the assertion generator associates equivalent Class 1 assertions both with line L1 of program, which calls main, and line L0 of main. Forward propagation of this latter assertion associates equivalent Class 1 assertions with L1 and L2 of main on the first pass through the loop, and similarly with L3 of main, because get_input does not change partition. But the generated Class 1 assertion at L3 refutes the Class 2 assertion stating sanitized(partition) associated with L3 (see above) generated from the specification of sanitize_partition. The lower left window of Figure 1 displays, as (LEMA) feedback to the developer, details of the resulting property violation, including a trace of the assertions that lead to the violation. Clicking on a line of LEMA code presents the user with a menu allowing additional feedback, e.g., a display of associated assertions and highlighting corresponding lines of C code.

5   Related Work

Most related work on application-specific security properties relates to data flow. The seminal work of Bergeretti et al. [6] provides rules for generating assertions about data flow and information flow in `while` programs. Amtoft et al. [2] automatically compute data flow assertions and contracts for SPARK Ada programs. The RESIN system, described in [19], facilitates developer specification of desired data flow properties; unlike in our approach, these specifications are used for runtime checking.

6   Progress and Future Work

Implemented to date are a C-to-LEMA translator for a significant subset of C with restricted use of pointers and loops, a database schema in MySQL [9], and theorem proving support using PVS. We also have a conceptual design for the GUI, plus designs for assertion generation techniques and for stored procedures to exercise the database during computation and analysis of the LEMA code. Future work includes 1) full integration of the prototype tools; 2) implementation of assertion generation from property specifications, code, and developer annotations; 3) support for handling pointers and unbounded loops; and 4) support for developer use of library routines.

Acknowledgements. We thank Prof. Wei Ding of the University of Massachusetts, Boston, who designed the database schema to fit our requirements for SecProve.

References

1. Coverity Prevent. Tool available at http://coverity.com/
2. Amtoft, T., Hatcliff, J., Rodríguez, E.: Precise and Automated Contract-Based Reasoning for Verification and Certification of Information Flow Properties of Programs with Arrays. In: Gordon, A.D. (ed.) ESOP 2010. LNCS, vol. 6012, pp. 43–63. Springer, Heidelberg (2010)
3. Archer, M., Leonard, E.: Establishing high confidence in code implementations of algorithms using formal verification of pseudocode. In: Proc. 3rd International Verification Workshop (VERIFY 2006), Seattle, WA (August 2006)
4. Archer, M., Leonard, E., Gasarch, C.: Verifying LEMA specifications in TAME. Technical Report NRL/MR/5540-12-9442, NRL, Wash., DC (to appear, 2012)
5. Barnett, M., Chang, B.-Y.E., DeLine, R., Jacobs, B., Leino, K.R.M.: Boogie: A Modular Reusable Verifier for Object-Oriented Programs. In: de Boer, F.S., Bonsangue, M.M., Graf, S., de Roever, W.-P. (eds.) FMCO 2005. LNCS, vol. 4111, pp. 364–387. Springer, Heidelberg (2006)
6. Bergeretti, J.-F., Carré, B.A.: Information-flow and data-flow analysis of while-programs. ACM Trans. Prog. Lang. Syst. 7(1), 37–61 (1985)
7. Bernardy, J.-P.: Lazy functional incremental parsing. In: Proc. of Second ACM SIGPLAN Symposium on Haskell (September 2009)
8. CodeSonar, http://www.grammatech.com/products/codesonar/overview.html

9. Ding, W., Archer, M.M., Leonard, E.I.: A database implementation to support program analysis. Draft report, NRL (2012)
10. Engler, D., Musuvathi, M.: Static Analysis versus Software Model Checking for Bug Finding. In: Steffen, B., Levi, G. (eds.) VMCAI 2004. LNCS, vol. 2937, pp. 191–210. Springer, Heidelberg (2004)
11. Flanagan, C., et al.: Extended static checking for Java. In: Proc. 2002 ACM SIGPLAN Conf. on Prog. Lang. Design and Implem. (PDLI 2002) (June 2002)
12. Gannod, G.C., Cheng, B.H.: Strongest postcondition semantics as the formal basis for reverse engineering. J. Automated Software Engineering 3(1/2) (1996)
13. Ghezzi, C., Mandrioli, D.: Incremental parsing. ACM Transactions on Programming Languages and Systems 1(1), 58–70 (1979)
14. Heitmeyer, C., Archer, M., Leonard, E., McLean, J.: Formal specification and verification of data separation in a separation kernel for an embedded system. In: Proc. 13th ACM Conf. on Comp. and Comm. Sec. (CCS) (2006)
15. Heitmeyer, C., Archer, M., Leonard, E., McLean, J.: Applying formal methods to a certifiably secure software system. IEEE Trans. Softw. Engin. (January/February 2008)
16. McGraw, G.: Software security: Building security in. Addison-Wesley (2006)
17. Pan, S., Dromey, R.G.: Using strongest postconditions to improve software quality. In: IFIP Conference Proceedings, vol. 3, pp. 235–240 (1994)
18. Shankar, N., et al.: PVS Prover Guide, Version 2.4. Technical report, Comp. Sci. Lab., SRI Intl., Menlo Park, CA (November 2001)
19. Yip, A., Wang, X., Zeldovich, N., Kaashoek, M.F.: Improving application security with data flow assertions. In: Proc. ACM SIGOPS 22nd Symp. Oper. Sys. Prin., SOSP 2009, pp. 291–304 (2009)

Anatomy of Exploit Kits*
Preliminary Analysis of Exploit Kits as Software Artefacts

Vadim Kotov and Fabio Massacci

DISI - University of Trento, Italy
surname@disi.unitn.it

Abstract. In this paper we report a preliminary analysis of the source code of over 30 different exploit kits which are the main tool behind drive-by-download attacks. The analysis shows that exploit kits make use of a very limited number of vulnerabilities and in a rather unsophisticated fashion. Their key strength is rather their ability to support "customers" in avoiding detection, monitoring traffic, and managing exploits.

Keywords: exploit kits, web threats, malware analysis.

1 Introduction

Over the last few years, the volume of web-borne malware significantly increased. According to various security reports [1,10] malicious URLs attacking browsers and their add-ons constitute the majority of all Internet threats. They exploit vulnerabilities in the web browsers and their add-ons in order to download malware executable onto the victim machine. This kind of attack is called drive-by-download [11]. In the worst cases compromised clients behind a company firewall can be used to wreak havoc on critical systems. In the best ones, they lay the basis for a large malware infrastructure that can be used for identity theft or banking fraud.

Drive-by downloads are managed by a so called *exploit kit* (or exploit pack) - a server application delivering malware instead of web content [12]. Its key feature is that in order to deploy it a "customer" of this tool does not need to be more expert in web technologies than a lousy system administrator. One only need to pay the developer of the kit for the code and possibly other services (such as obfuscation). These characteristics ultimately increase the number of possible attackers and the risks for the community at large.

1.1 Our Goals and Contribution

In this paper we explore the leaked source code for some popular exploit kits. In our analysis we pursued the following goals:

---

* Work partly supported by the European Union by the Erasmus Mundus Action 2 Programme and the Project EU-FP7-SEC-CP-SECONOMICS.

J. Jürjens, B. Livshits, and R. Scandariato (Eds.): ESSoS 2013, LNCS 7781, pp. 181–196, 2013.

- Study the functional aspects of exploit kits and offer a taxonomy for the routines implemented in them;
- Classify the exploit delivery mechanisms;
- Uncover web crawler evasion techniques that are used by exploit kits.
- Understand the user interface of an exploit kit, find out what data it provides and what management capabilities are available to the customer.
- Investigate the code re-use in various exploit kits and determine if there is a common code base used by malware authors.
- Study the methods of code protection mechanisms that are aimed to prevent unauthorized code distribution and complicate the analysts' work.

The results of our study are quite surprising. We expected exploit delivery mechanisms to be sophisticated - to work as snipers, performing a careful study of the remote machine and then delivering only the right exploit to the right victim. While the study is performed by most kits, its results are not used in a significant way to select the exploit. Instead the attack is performed in machine-gun style. It seems that the main purpose of victim fingerprinting is to make statistics and "dashboards" of traffic and malware installations. In other words exploit kits' main target is to "improve the customer experience". A large number of successfull infections is expected to come by large volumes of traffic instead of sophisticated attacks.

2   Related Works

Very few papers have examined exploit kits as a class of software artefacts. Most studies on infiltrations (such as those by Savage, Paxson and their groups [4,9]) usually focus on a single tool and try to reconstruct the whole food chain from the web-user to the final bad guy monetizing the result. For example, Motoyama et al. [9] analyzed the private messages exchanged in 6 underground forums. They analyzed whether sellers did re-use the same ID, whether transactions were moderated, or reputation systems were in place. A similar study focusing on the Chinese sites has been done in [13]. Yet they did not consider analyzing the actual malware posted on those forums. Franklin et al. [4] and Herley et al. [8] have analyzed (with opposite conclusions) the whole chain for spam and malware goods distribution but have not considered the individual artefacts at the start of the chain. Grier et al. [6] described the landscape of exploit kits and malware families, with more detailed focus on the latter. Their main result is a statistical analysis that shows which exploit kits are used to distribute which malware on what kind of traffic. For example, the authors determined that modern exploit kits deliver 32 different malware families including, ZeroAccess, SpyEye and Zeus as ones of the most famous. But there were no analysis of exploit kit technologies as much.

Only the author of [7] did study exploitation capabilities of the popular malware toolkits. However, the paper only considers a small number of instances and does not provide a comparison of their features as software artefacts. Our

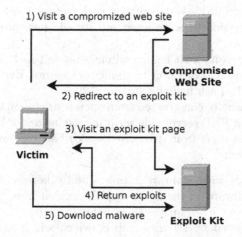

1) Visit a compromized web site

Compromised Web Site

2) Redirect to an exploit kit

Victim

3) Visit an exploit kit page

4) Return exploits

5) Download malware   **Exploit Kit**

Fig. 1. Scheme of drive-by-download attack

perspective is to take a wider look and investigate the structure of the crimeware packs. Another paper in which the actual instances have been considered is the work by Cova et al.[3] which focuses on fishing kits.

3 Background

An exploit kit is a software tool traded on the black market and used by cybercriminals to perform drive-by-download attacks. From an implementation point of view an exploit kit is an HTTP server-side application, that, based on request headers, returns a page with an appropriate set of exploits. Its main purpose is silently downloading and executing malware on the victim machine by taking advantage of browser vulnerabilities. Errors in applied programming interfaces or memory corruption based vulnerabilities allow an exploit to inject a set of instructions (called shellcode) into the victim process. Shellcode on its turn downloads a malware executable to the victim's hard drive and executes it. The executable that gets installed on the target system is completely independent from the exploit pack (see [6] for a distribution of malware families provisioned by the different exploit kits). An owner can "arm" it with any malicious application of her choice.

Fig. 1 depicts the generic scenario of drive-by-download attack [11]. A victim visits a compromised web site, from which she gets redirected to the exploit kit page. Various ways of redirection are possible: an <iframe> tag, a JavaScript based page redirect etc. The malicious web page then returns an HTML document, containing exploits, which are usually hidden in an obfuscated JavaScript code. If at least one exploit succeeds, then a victim gets infected. Successful exploitation means, that the shellcode injected has finished flawlessly and hence accomplished its task - to download and execute a malicious program.

How successful an exploit kit is depends on such factors as an operating system version, type and version of a browser and its add-ons, presence of security measures.

Apart from the exploits a kit has an administrative panel - a dashboard that provides statistics and allows a user to configure the tool. Even the earliest kits such as Mpack and IcePack had this feature [12].

Exploit kits are usually constructed from open-source components such as an Apache web server, a PHP server-side scripting engine and a MySQL database.

In order to protect users from drive-by-download attack, two main strategies are normally deployed:

1. Protect end users with malware scanners and other security means which intercept the malware on the fly or stop the exploit from completing;
2. Build black lists of URLs (such as those behind Google Safe Browsing). These lists are constructed by security web crawlers, which instead of indexing the site content, check the web page with malware scanners or analyze its behavior in a sandbox. In fact this can be done by the search engine's robots in addition to traditional content indexing.

These two defence mechanisms determine the presence of yet another feature of an exploit kit: detection evasion. In this sense a kit can implement the following self-protection measures:

- Code obfuscation, deployed in order to fail malware scanners' signatures and heuristics. For example the Black Hole exploit kit [5] applies a polymorphic obfuscation algorithm to its malicious JavaScript code.
- Checking itself with antiviruses to find out whether the signature for the current obfuscation scheme already exists and whether it is time to update the obfuscation algorithm.
- Restricting search robots activity by disallowing indexing policy in the "robots.txt" file.
- Mimicking an innocent web page when encountering an unsupported user agent (search robots, downloading software, etc.).
- Looking itself up in the black lists of URLs and IP addresses (like Google Safe Browsing) and, if found, rebind itself to another server/domain.

Finally, an exploit kit is a software product in itself and therefore must have some features of legitimate software such as source code protection, licensing, binding to single server/domain, etc. For example the Fragus exploit kit is protected with a commercial tool IonCube[1], which also makes it impossible to run the kit under the domain/server [2] that is different from the customer's.

To better understand the idea of exploit kit let's consider the following example: a user running Firefox 1.0.4 with Adobe Reader v.8.1.1 under the Windows XP opens a web page from a compromised server. An invisible iframe (left by the hacker) loads a page from another web server hosting the Eleonore 1.4.4mod exploit kit. On the server side a corresponding PHP script parses the client's

---

[1] www.ioncube.com, checked on 14 Aug 2012

HTTP request headers and retrieves the following information: name and version of the browser (Firefox 4.0), name and version of operating system (Windows XP). Based on that the PHP script selects the set of exploits such as one for CVE-2005-2265 vulnerability (targeting Firefox 1.0.4). The exploits selected are wrapped in the JavaScript code, which then gets obfuscated and returned to the client. If the exploit succeeds (if nothing stops it from executing), then the shellcode takes control over the browser's execution. It calls the URLDownload-ToFile (Urlmon.dll) and then WinExec (Kernel32.dll) functions to download and execute an instance of Zeus trojan that is stored on the attacker's web server. Once the shellcode makes the request to the exploit kit server (to retrieve the trojan), the corresponding PHP script adds the successful exploitation record to the database and returns the contents of the binary. The owner of this malicious server in the administrative panel can see how many visitors were lured to the malicious page and how many of them were infected.

The features that we have listed above are the capabilities of an exploit kit that could be implemented. Whether they are really used in real-world tools and if yes then to what extent - is the question that we try to answer in this paper.

4   Collected Data

To collect the data for analysis two sources of information were used:

- A list of exploit packs, available at Contagio Malware Dump [2] security blog.
- Advertising and leaked code on various black hat forums.

Altogether we identified information for more than 70 exploit kits and out of those we were able to successfully deploy 33 instances of 24 families. Our semantics of successful is that the kit installs, runs, and is able to deliver a prototype malware of our choice to an appropriate client. We are now running a more sophisticated experiment in which we benchmark whether all claims about number of successful installations by the exploit kit developer in terms of successful installation are correct. The full list of deployed kits is presented in Table 2. Our collection includes the most famous products on the black market, according to the reports of Kaspersky Lab [12], Sophos [5] and Symantec [2].

Among all deployed exploit kits there is one that we can not yet fully analyse - Crimepack v.3.1.3. It was obfuscated with a powerful commercial protector named IonCube for which, to our knowledge, there is no good deobfuscation tool. But we were still able to extract some information from Crimepack using black box analysis of the deployed sample.

Figure 2 shows the connections between an exploit kit and some related entities such as the victim, the malware scanner or security crawler and the developer.

All the kits in our collection were written in PHP and were designed to work in bundle with MySQL database. They present the following key architectural components:

---
[2] http://contagiodump.blogspot.it/2010/06/
overview-of-exploit-packs-update.html, checked on 14 Aug 2012.

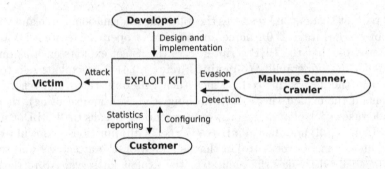

Fig. 2. Exploit kit use cases

Offensive Component which is responsible for analyzing and ultimately attacking vulnerable machines;

Defensive Component that protects the toolkit from detection by malware scanner (such as obfuscation of pages);

Management Component which supplies the reporting and configuration components of an exploit kit to support the customer;

Protection Mechanisms , which includes means of protection applied to an exploit kit for preventing unauthorized distribution and complicating the process of reverse engineering.

5   Offensive Component

To identify the operation of the offensive routes we performed code inspection, debugging and sandboxed code execution. The offensive routine consists mainly of two parts. The first one occurs on the request of the web client, when a victim gets redirected to the bad page. If the exploit is successful, the second part is activated on shellcode request after it has taken control over the target application.

The first part consists of the following steps:

1. *User agent detection* determines operating system (OS) and user agent (UA) used by the victim.
2. *IP blocking* blocks a visitor on the next visit (based on IP address).
3. *UA validation* - If the OS or the UA are not supported do either of the following actions:
   - output an innocent looking page like "Site is under construction", and provide the response status 200 (OK);
   - redirect a visitor to another page or web site by specifying its address in the "Location" header of the server response;
   - output an error page and provide the response status of error (e.g. 404);
   - deliver anyhow some exploits in the hope that the client is vulnerable.
4. *Exploits selection* - Select the subset of exploits for the determined OS and UA or follow corresponding execution branch.

Table 1. Presence of the offensive routine steps

Step	Present (%)	Absent (%)
User agent detection	100%	0
IP blocking	79%	21%
UA Validation	88%	12%
Exploits selection	82%	15%
Exploits obfuscation	82%	18%
Executable Delivery	100%	0%

5. *Exploits obfuscation* - Obfuscate the generated malformed HTML page. By this we mean the on-line obfuscation, when the attacking page goes through some transformations that change its appearance.

The second part has only one step, which is:

1. *Delivery of malware executable* - returns the malware executable file and as a follow up updates the successful exploitation statistics.

Each exploit kit largely follows the proposed scenario, irrespective of the year of deployment: Icepack kit appeared in 2007, while Phoenix is a comparatively recent product, its 3.1 update was released in 2012. Therefore we conjecture that the functional architecture is essentially stable.

A summary of the findings is shown in Table 1. The full analysis of the server side attack scenario can be found in Table 2. The results do not sum up to 100% as in some cases it was not possible to ascertain exactly what the kit does. Out of these results we can make some conclusions:

1. 88% of exploit kits perform user agent validation, which means that if a browsing robot wants to detect an attack it must send a user agent string of a vulnerable browser (Internet Explorer 6 under Windows XP is going to work for every kit analyzed) or, on the other hand, a user can change the user agent string to an unsupported one (e.g. wget under OpenBSD) in order to "trick" the kit and avoid infection.
2. 64% of exploit kits perform both IP blocking and Exploits selection, which complicates the analysis of a kit in the wild. Offensive capabilities of an exploit kit in the wild can only be revealed from different IP addresses and using different user agent strings.
3. All exploit kits in our collection have a separate piece of code responsible for executable delivery. Thus, exploit kits can keep an accurate score of the machines that were actually infected.
4. In a surprisingly large number of cases (36%), irrespective of the result of the UA validation, the exploit kit will anyhow throw some attacks. The UA validation code does not seem to be used in a significant way to select the exploit appropriately.

In terms of vulnerability analysis the picture was surprising: among the 70+ exploit kits that we had identified only a bit more than 110 vulnerabilities are

Table 2. Full data set and offensive component analysis results

Name	Version	IP Block.	UA Detec.	UA Valid.	Follow-up	Sel.	Obf.
0x88	UNK	✓	✓	✓	ATTACK	✓	✓
adpack	UNK1	✓	✓	✓	INNOCENT	✓	✓
adpack	UNK2	✓	✓		NONE		
armitage	1.0 beta	✓	✓	✓	INNOCENT	✓	✓
bleeding life	2		✓	✓	INNOCENT	✓	
crimepack	3.1.3	✓	✓	✓	INNOCENT	UNK	✓
cry217	UNK	✓	✓		NONE		
eleonore	1.2	✓	✓	✓	ATTACK	✓	✓
eleonore	1.4.4 mod	✓	✓	✓	ATTACK	✓	✓
firepack	0.18	✓	✓	✓	ERROR	✓	✓
firepack	UNK	✓	✓	✓	INNOCENT	✓	✓
fragus	1.0	✓	✓	✓	ATTACK	✓	✓
fragus	black	✓	✓	✓	ATTACK	✓	✓
gpack	UNK	✓	✓	✓	INNOCENT	✓	✓
icepack	platinum beta	✓	✓	✓	EMPTY	✓	✓
icepack	platinum	✓	✓	✓	INNOCENT	✓	✓
el fiesta	1.0		✓	✓	INNOCENT	✓	✓
el fiesta	1.8	✓	✓	✓	INNOCENT	✓	✓
life	UNK	✓	✓		NONE		
mpack	0.81	✓	✓	✓	INNOCENT	✓	✓
mpack	0.86	✓	✓	✓	INNOCENT	✓	✓
mpack	0.91	✓	✓	✓	INNOCENT	✓	✓
mpack	0.99	✓	✓	✓	INNOCENT	✓	✓
mypolysploit	1.0		✓	✓	ATTACK	✓	✓
neon	UNK	✓	✓	✓	ATTACK	✓	✓
nuke	UNK		✓	✓	INNOCENT	✓	✓
phoenix	2.3	✓	✓	✓	INNOCENT	✓	
rds	2.0	✓	✓		NONE		✓
salo	UNK		✓	✓	ATTACK	✓	
seo	UNK		✓	✓	INNOCENT	✓	✓
shaman's dream	2.0	✓	✓	✓	ATTACK	✓	✓
unique	UNK	✓	✓	✓	INNOCENT	✓	✓
yes	2.0		✓	✓	REDIRECT	✓	✓

Explanation of the table columns:

Name - name of an exploit kit;
Version - exploit kit version or UNK if we could not determine it;
IP Block. - presence if IP blocking: YES (✓) or NO.
UA detec. - detection of the user agent: YES (✓) or NO.
UA valid. - user agent validation, i.e. an action taken if a user agent is not supported: INNOCENT | REDIRECT | ERROR | ATTACK | NONE, where INNOCENT means an innocent looking page; REDIRECT - a redirection provided is "Location" header; ERROR - an error page; ATTACK - throw some exploits; NONE - if no action is taken.
Sel. - presence of exploit selection: YES(✓) if, based on user agent information, execution path of the kit changes, otherwise NO.
Obf. - presence of exploit obfuscation: YES(✓) or NO.

actually exploited. An average exploit kit had around 10 exploits ($\mu = 11.1$) which are not always fresh. Table 3 shows the mean number of exploits of certain age over the sample of 30 kits. Age of an exploit was calculated relatively to the year this kit first appeared. On average, most exploits are aimed at 1 and 2 years old vulnerabilities, which may imply that malware authors prefer to use public exploits, rather than private ones. An alternative explanation is that the time to market a reliable piece of code exploiting commodity software is significant.

The affected software is also very limited, showing a preference among exploit kit developers for easy exploits based on popular software. Figure 3 shows how many vulnerabilities affecting a given software are present in the overall sample.

Table 3. Average number of exploits by age

6 y.o.	5 y.o.	4 y.o.	3 y.o.	2 y.o.	1 y.o.	0 y.o.
0.17	1.03	1.4	2.1	2.57	3.93	1.9

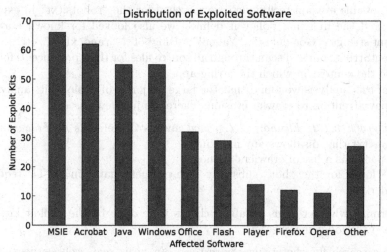

The figure shows a number of exploit kits (vertical axis) that exploit particular software or class of software (horizontal axis). *Player* denotes various
video/audio players (such as Real Player, Quick Time etc.), *Windows* includes
exploits for components of Windows operating system, *Other* denotes various
other types of exploited software (such as components of Microsoft Visual Studio, audio/video format converters, messengers etc).

Fig. 3. Preferred Software for Exploits

The entry 'Player' denotes various proprietary or open source video/audio
players (such as Real Player, Quick Time etc.), while 'Windows' includes exploits for various components of Windows operating system of different versions. The category 'Other' denotes various other types of exploited software
(such as components of Microsoft Visual Studio, audio/video format converters,
messengers etc).

5.1   Defensive Component

Defensive means of exploit kits include IP blocking, payload obfuscation, crawlers
evasion and active measures such as checking itself in various virus databases
to catch the time, when it got recognized by the malware scanners in order to
update the obfuscation scheme.

As we mentioned, IP blocking and obfuscation are popular measures routinely
deployed in the offensive component.

Crawlers evasion can be implemented in two ways (not mutually exclusive):

1. Settings the specific indexing policy in a "robots.txt" file, which may keep search robots from collecting the information about malicious pages of an exploit kit;
2. Match a user agent string in HTTP request against known crawlers.

To find possible evasion techniques, we searched for the "robots.txt" in exploit kit files and the indexing policy it defines; we also looked for known crawler user agent strings ("Googlebot", "Yahoo!", "Bingbot", "YandexBot", etc.) from UserAgentString.com and scan through all source files for their presence. If found - analyse the context in which the string appears.

Out of this analysis we found that the large majority of exploit kits analyzed do not pay attention to crawler evasion. There are just few cases:

1. *Crimepack 3.1.3, Eleanore 1.4.4 mod* and both versions of *Fragus* have robots.txt that disallows any indexing.
2. Firepack has a list of crawlers' names.
3. *0x88* looks for the "bot" substring in user agent name in HTTP request header.

To determine whether virus database checks were done by the exploit kit, we performed the following checks:

1. String search for the strings "virustotal" (a virus scan web service[3]) and "virtest" (a popular anonymous virus scan web service[4]), that can locate the code snippets responsible for virus database checks.
2. Looking at administrative panels of exploit kit to find the pieces of user interface that might indicate the presence of the virus database checks.

No exploit kit (except Crimepack 3.1.3 for which it is unknown) checks itself in the virus databases. However, we have not examined the source of Black Hole exploit kit. So we cannot confirm [5], which reports that it has the ability to check itself against two virus scan services.

Figure 4 shows a Venn diagram where the explicit number of items in each subset is represented by the number of crosses. Whether a kit may use or not the IP Blocking, an overwhelming majority uses obfuscation. Therefore the absence of obfuscation in a page seems a good indication that the page is unlikely to be malicious.

6   Management Component

The customer-oriented part of an exploit kit should provide an access to visits and exploitation statistics and offer some settings to manage the toolkit.

A step-by-step work flow can not be proposed here, because the customer handles the exploit kit by an interactive user interface. The main *use cases* are the following:

---

[3] http://www.virustotal.com, checked on 14 Aug 2012
[4] http://www.virtest.com, checked on 14 Aug 2012

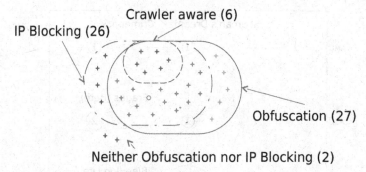

IP Blocking (26)

Crawler aware (6)

Obfuscation (27)

Neither Obfuscation nor IP Blocking (2)

The figure shows a Venn diagram of the defences implemented in the exploit kits analyzed. The number of items in each subset is represented by the number of crosses.

Fig. 4. Venn diagram of defensive capabilities

1. *Installation* - Install the exploit kit, i.e. allocate all resources needed for successful run.
2. *Authentication* - Authenticate exploit kit user.
3. *Control Cockpit* which normally includes
   - site visit statistics;
   - successfull exploitation statistics;
   - exploit kit settings;

Since we dealt with *leaked* code, the analysis of the installation step can only be preliminary, because automated installers might have been removed after setting up the kit. Therefore we can not say whether an automatic installer supposed to come with the software or not.

Authentication is present in every admin panel, so we do not discuss it further in detail.

The two important functions of customer oriented component are statistics reporting and settings. All kits feature some basic setting up and statistics, and a more fine grained classification is reported below:

1. *Toolkit statistics* includes the information about the exploit kit's work: total visits, exploited systems, browsers and operating systems.
2. *Market statistics* includes the information that helps a customer to manage her interaction with related markets (e.g. traffic market). The typical information of this type are referers[5] and/or countries.
3. *Exploit statistics* allows a customer to track the effectiveness of individual exploits.

We have similar three grades metric for the settings, that can be offered to the customer:

---

[5] A URL from which the victim came. It can be determined from the HTTP header "Referer".

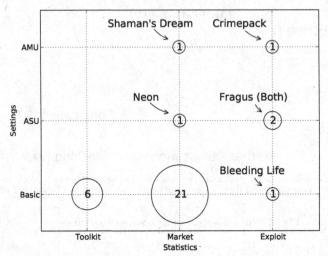

The figure shows a magic quadrant depicting the exploit kits in the statistics/settings coordinate axes.
The grades for the settings axis:

Basic allows a customer to change her credentials and replace malware executable

ASU (Advanced single-user gives a customer more flexibility (manage exploits, build black lists by country, referer etc.)

AMU (Advanced multi-user) provides the settings for multi-user environment.

The grades for statistics axis:

Toolkit includes the information about the exploit kit's work: total visits, exploited systems and browsers etc.

Market statistics includes the information that helps a customer to manage her interaction with related markets (e.g. traffic market).

Exploit statistics allows a customer to track the effectiveness of individual exploits.

Fig. 5. Statistics/Settings Quadrant

1. *Basic settings* - allows a customer to change her credentials and replace malware executable.
2. *Advanced single-user settings (ASU)* - allows a customer to enable/disable exploits or/and add new ones or/and manage block lists (IP, referrers, countries) etc.
3. *Advanced multi-user settings (AMU)* - provides the settings for multi-user environment, i.e. customizable user profiles.

To perform this analysis we need to have a look at every administrative panel of our exploit kit collection and enumerate the reported statistics and the configuration options that are offered to the customer.

In Figure 5 we provide a magic quadrant depicting the exploit kits in the statistics/settings coordinate axes. The number inside the circle is a count of exploit kits that fell upon the corresponding point in the quadrant.

It can be seen that the majority of exploit kits analyzed produce advanced statistics, but offer the simplest settings possible. From statistics/settings perspective Crimepack, Fragus and Shaman's Dream are the most advanced. The conjecture is that the closer an exploit kit is to the top right corner of the quadrant the higher is its potential of supporting complex business models. In other words the advanced statistics and settings allow customers to

- keep track of the traffic they actually received, possibly to match it against what they may have bought in order to avoid waste of money in useless traffic,
- learn which exploits contribute more to the victim infecting process in order disable the bad ones (or detectable ones) and so forth.

This gives a customer a high level of flexibility in organizing the infection process and ways of its monetizing. Top right corner is the place where the Black Hole exploit kit would have been placed.

7 Code Protection

Presence of protection in an exploit kit source code can be detected based on one of the two features: (1) there is a byte code and markers of the commercial tools (Zend Guard[6] or IonCube) or (2) the code is put through some permutations and then executed. Otherwise the code is clear to read.

In summary our findings are reported below:

- *Crimepack 3.1.3* - is the only kit in our collection that uses IonCube.
- *Neon*, *Life* and *Firepack$_{UNK}$* use Zend Guard.
- *0x88*, *Eleonore 1.4.4 mod*, *El Fiesta 1.0* and *Unique* use various ad-hoc methods of protection.
- Other kits (25) do not use any code protection.

Yet these results can not be taken as conclusive because we were dealing with *leaked* sources. For example, Fragus exploit kit, according to [2] is obfuscated with IonCube, while we obtained a clean version. All conclusions that are made from this analysis can give only general idea of the code protection within exploit kits. One of the reason the Black Hole exploit kit is not included in the study is that we could not fully restore the source code.

8 Code Re-use

Investigating the cases of code re-use could help us to better understand the production of an exploit kit. To address this question we use a token-based copy-paste detector phpcpd[7]. It reveals the snippets of repeating code among multiple PHP scripts.

---

[6] http://www.zend.com/en/products/guard/, checked on 14 Aug 2012
[7] https://github.com/sebastianbergmann/phpcpd/, checked on 14 Aug 2012

Table 4. Numbers and functionality of code re-use cases per pair

Kit$_1$	Kit$_2$	# of matches	Repeated Code Functionality
0x88	life	7	Admin. panel routines, obfuscation, database functions
fragus	icepack	5	Obfuscation
adpack	gpack	3	User agent detection, database functions
armitage	icepack	3	User agent detection
mpack	0x88	3	Obfuscation
rds	0x88	3	Admin. panel routines, obfuscation, database functions
eleonore	shaman	2	User agent detection, obfuscation
firepack	icepack	2	Obfuscation
nuke	seo	2	Admin. panel routines, user agent detection
yes	icepack	2	User agent detection
lefiesta	fragus	1	Obfuscation
neon	armitage	1	Array of countries names
salo	adpack	1	Obfuscated array of countries names

The detected re-used code can be divided into three groups:

1. the code repeated among different versions of the same exploit kit;
2. the code, that corresponds to open source libraries, that are used by different kits;
3. the code repeated in different *families* of exploit kits.

The first group of code is not interesting because the results of the analysis were predictable. The kits of the same family have a lot of common code.

In the second group the only open source PHP library that we have found in our collection was Geo IP[8] which allows to determine the country by IP address. This library is frequently used in exploit kits to provide country related statistics.

The third group of repeated code snippets consists of the those appearing in the different exploit kit families. The summary of code re-use cases between *different families of exploit kits* is shown in Table 4.

The highest volumes of "copy-paste" can be found at (0x88, life) and (fragus, icepack) pairs. Interestingly there is a code obfuscation algorithm that was implemented in 5 different kits (Mpack, Fragus, RDS, Life and 0x88).

In our collection there are total of 24 exploit kit families, since the number of possible pairs is $C_{24}^2 = \frac{24!}{2!(24-2)!} = 276$ and the number of pairs with at least one repeated snippet is 13, then the rate of code re-use (based on token analysis) is $13/276 = 0.047$. This means that based on analysis of the PHP language tokens similarity *there is no common code base that is used by the malware authors*. We can use the above observation to conclude that the market is also fragmented with multiple kit providers.

9   Conclusion

In this paper we have reported the first analysis of exploit kits as software artefacts. We have collected information on 70+ popular tools for malware

---

[8] http://php.net/manual/en/book.geoip.php, checked on 14 Aug 2012

distribution. Out of those we have been able to successfully deploy and test 30+ kits. They have been further analyzed.

In order to understand the nature of the class we have collected a set of leaked source code files of previously private exploit kits. Each of them we have tried to deploy, and those that we succeeded to run were selected for analysis. We used a combination of analysis techniques such as static and dynamic reverse engineering.

The result of the work can be summarized as follows:

- Exploit kits have very similar functionality, largely following the work flow described in this paper. The victim is fingerprinted (user agent and operating system information together with IP address are collected partly for exploit selection and partly for statistical purposed), a set of moderately old exploits is provisioned within an obfuscated web page in order to download and execute a malware program. Very few vulnerabilities are exploited.
- Exploit kits use IP blocking, user agent validation and code obfuscation. Very rarely they try to evade web crawlers.
- Most exploit kits provide customers with statistics and settings, allowing them to customize a toolkit and track its activity. A customer can use them to interact with other types of black markets - traffic, malware executables, hosting services, etc.
- Results of token based copy-paste analysis show that the kits analyzed seem to be written mostly independently one from another, without a common code base.
- Some exploit kits use commercial code protection (e.g. Crimepack), which means that malware authors expect to get significant amounts of money from the crimeware toolkit sales.

We expect the exploit kit technology to evolve further in direction of detection evasion and enhancement of the customer's experience. The evidence of this we can see already in the latest kits, such as Black Hole v.2. The little emphasis on exploit delivery seems to imply that a better protection can stem from large scale detection rather than individual protection.

References

1. Internet security threat report (April 2012),
   http://www.symantec.com/threatreport (Checked on September 10, 2012)
2. Coogan, P.: Fragus exploit kit changes the business model (February 2010),
   http://www.symantec.com/connect/blogs/
   fragus-exploit-kit-changes-business-model
   (Checked on September 10, 2012)
3. Cova, M., Kruegel, C., Vigna, G.: There is no free phish: an analysis of 'free' and live phishing kits. In: Proceedings of WOOT 2008, pp. 4:1–4:8 (2008)
4. Franklin, J., Paxson, V., Perrig, A., Savage, S.: An inquiry into the nature and causes of the wealth of internet miscreants. In: Proceedings of CCS 2007, pp. 375–388 (2007)

5. Fraser, H.: Exploring black hole exploit kit (March 2012),
   http://nakedsecurity.sophos.com/exploring-the-blackhole-exploit-kit
   (Checked on September 10, 2012)
6. Grier, C., Ballard, L., Caballero, J., Chachra, N., Dietrich, C.J., Levchenko, K.,
   Mavrommatis, P., McCoy, D., Nappa, A., Pitsillidis, A., Provos, N., Rafique, M.Z.,
   Rajab, M.A., Rossow, C., Thomas, K., Paxson, V., Savage, S., Voelker, G.M.:
   Manufacturing compromise: the emergence of exploit-as-a-service. In: Proceedings
   of the 2012 ACM Conference on Computer and Communications Security, CCS
   2012, pp. 821–832. ACM, New York (2012)
7. Guido, D.: A case study of intelligence-driven defense. IEEE Security Privacy 9(6),
   67–70 (2011)
8. Herley, C., Florencio, D.: Nobody sells gold for the price of silver: Dishonesty,
   uncertainty and the underground economy. In: Economics of Information Security
   and Privacy (2010)
9. Motoyama, M., McCoy, D., Savage, S., Voelker, G.M.: An analysis of underground
   forums. In: Proceedings of ICM 2011 (2011)
10. Namestnikov, Y.: IT threat evolution: Q1 2012 (May 2012),
    http://www.securelist.com/en/analysis/204792231/
    IT_Threat_Evolution_Q1_2012 (Checked on September 10, 2012)
11. Naranie, R.: Drive-by downloads. The web under siege (April 2009) (Checked on
    September 10, 2012)
12. Preuss, M., Diaz, V.: Exploit kits - a different view (February 2011),
    http://www.securelist.com/en/analysis/204792160/
    Exploit_Kits_A_Different_View(Checked on September 10, 2012)
13. Zhuge, J., Holz, T., Song, C., Guo, J., Han, X., Zou, W.: Studying malicious
    websites and the underground economy on the chinese web. In: Proceedings of
    MIRES, pp. 225–244 (2009)

An Empirical Study on the Effectiveness of Security Code Review

Anne Edmundson[1], Brian Holtkamp[2], Emanuel Rivera[3],
Matthew Finifter[4], Adrian Mettler[4], and David Wagner[4]

[1] Cornell University, Ithaca, NY, USA
[2] University of Houston–Downtown, Houston, TX, USA
[3] Polytechnic University of Puerto Rico, San Juan, Puerto Rico
[4] University of California, Berkeley, CA, USA

Abstract. With the rise of the web as a dominant application platform, web security vulnerabilities are of increasing concern. Ideally, the web application development process would detect and correct these vulnerabilities before they are released to the public. This research aims to quantify the effectiveness of software developers at security code review as well as determine the variation in effectiveness among web developers. We hired 30 developers to conduct a manual code review of a small web application. The web application supplied to developers had seven known vulnerabilities, including three different types: Cross-Site Scripting, Cross-Site Request Forgery, and SQL Injection. Our findings include: (1) none of the subjects found all confirmed vulnerabilities, (2) more experience does not necessarily mean that the reviewer will be more accurate or effective, and (3) reports of false vulnerabilities were significantly correlated with reports of valid vulnerabilities.

1 Introduction

With the widespread adoption of and reliance on the Internet, web applications are playing an increasing role in our everyday life. With such a large user base, web applications have become a prime target for attackers who wish to hijack websites or to steal user information. Unfortunately, it is common for these applications to be susceptible to attacks. Web application vulnerabilities primarily result from bugs in application-specific code. These arise due to a widespread lack of expertise about web security among developers, and frequently involve departures from coding best practices.

Ideally, web applications would be free of vulnerabilities and therefore secure. While it is difficult to determine whether any vulnerabilities remain in an application, it is generally believed that an application is more secure when it has fewer vulnerabilities. It is therefore common for software developers and companies to make efforts to find and eliminate vulnerabilities in their software. Two common ways of accomplishing this are by manually reviewing source code and by using automated tools that are capable of identifying vulnerabilities.

In this study we focus on one of these techniques: manual code review. Specifically, we aim to measure the effectiveness of manual code review of web applications for improving their security. We used a labor outsourcing site to hire 30 web developers

J. Jürjens, B. Livshits, and R. Scandariato (Eds.): ESSoS 2013, LNCS 7781, pp. 197–212, 2013.

with varying amounts of security experience to conduct a security code review of a simple web application. These developers were asked to perform a line-by-line code review of the application and submit a report of all security vulnerabilities found. Using the data we collected:

- We quantified the effectiveness of developers at security code review.
- We estimated the optimal number of independent reviewers to hire to achieve a desired degree of confidence that all bugs will be found.
- We measured the extent to which developer demographic information and experience can be used to predict effectiveness at security code review.

These results may help hiring managers and developers in determining how to best allocate resources when securing their web applications.

2   Goals

In this work, we conduct an exploratory analysis of software developers' effectiveness in conducting security code review. We are interested in determining: (1) how effective developers are at conducting code reviews and the degree of variation among them, (2) the optimal number of independent reviewers to hire, and (3) whether we can predict in advance which developers will be most effective at performing a security review.

2.1   Effectiveness

Our research measures how well developers conduct security code review. In particular, we are interested in how effective they are at finding exploitable vulnerabilities in a PHP web application, and how much the effectiveness varies between reviewers. We are interested in answering the following questions:

1. What fraction of the vulnerabilities can we expect to be found by a single security reviewer?
2. Are some reviewers significantly more effective than others?
3. How much variation is there between reviewers?

2.2   Optimal Number of Reviewers

Depending on the distribution of reviewer effectiveness, we want to determine the best number of reviewers to hire. Intuitively, if more reviewers are hired, then a larger percentage of vulnerabilities will be found, but we want to determine the point at which an additional reviewer is unlikely to uncover any additional vulnerabilities. This would be useful for determining the best allocation of resources (money) in the development of a web application. Specifically, we will address the following questions to find the optimal number of reviewers.

4. Will multiple independent code reviewers be significantly more effective than a single reviewer?
5. If so, how much more effective?
6. How many reviewers are needed to find most or all of the bugs in a web application?

2.3 Predicting Effectiveness

We asked each reviewer about the following factors, which we hoped might be associated with reviewer effectiveness:

– Application comprehension
– Self-assessed confidence in the review
– Education level
– Experience with code reviews
– Name and number of security-related certifications
– Experience in software/web development and computer security
– Confidence as a software/web developer and as a security expert
– Most familiar programming languages

Identifying the relationship between reviewers' responses to these questions and their success at finding bugs during code review may provide insight into what criteria or factors would be most predictive of a successful security review.

3 Experimental Methodology

To assess developer effectiveness at security code review, we first reviewed a single web application for security vulnerabilities. We then hired 30 developers through an outsourcing site and asked each of them to perform a manual line-by-line security review of the code. After developers completed their reviews, we asked them to tell us about their experience and qualifications. Finally, we counted how many of the known vulnerabilities they found.

3.1 Anchor CMS

We used an existing open-source web application for the review, Anchor CMS. Anchor CMS is written in PHP and JavaScript and uses a MySQL database. We chose this application for our study due to (1) the presence of known vulnerabilities in the code, (2) its size, which was substantial enough to be nontrivial but small enough to allow security review at a reasonable cost, and (3) its relatively permissive license, which let us anonymize the code, as described below.

There are currently four release versions of Anchor. We chose to have reviewers review the third release, version 0.6, instead of the latest version. This version had more known vulnerabilities while still having comparable functionality to the latest version.

To prepare and anonymize the code for review, we modified the Anchor CMS source code in two ways. First, we removed the Anchor name and all branding. We renamed it TestCMS, a generic name that wouldn't be searchable online. We did not want developers to view Anchor CMS's bug tracker or any publicly reported vulnerabilities; we wanted to ensure they reviewed the code from scratch with no preconceptions. Our anonymization included removal of "Anchor" from page titles, all relevant images and logos, and all instances of Anchor in variable names or comments.

Once the code was anonymized, we modified the code in two ways to increase the number of vulnerabilities in it. This was done in order to decrease the role of random noise in our measurements of reviewer effectiveness and to increase the diversity of vulnerability types. First, we took one vulnerability from a prior release of Anchor (version 0.5) and forward-ported it into our code. After this modification, the web application had three Cross-Site Scripting vulnerabilities known to us and no Cross-Site Request Forgery protection throughout the application.

Second, we carefully introduced two SQL injection vulnerabilities. To ensure these were representative of real SQL injection vulnerabilities naturally seen in the wild, we found similarly structured CMS applications on security listing websites (SecList.org), analyzed them, identified two SQL injection vulnerabilities in them, adapted the vulnerable code for Anchor, and introduced these vulnerabilities into TestCMS. The result is a web application with six known vulnerabilities. Our procedures were designed to ensure that these vulnerabilities are reasonably representative of the issues present in other web applications.

These six known vulnerabilities are exploitable by any visitor to the web application; he need not be a registered user. Additionally, the vulnerabilities are due solely to bugs in the PHP source code of the application. For example, we do not consider problems such as Denial of Service attacks or insecure password policies to be exploitable vulnerabilities in this study. Although these issues were not included in our list of six known vulnerabilities, we did not classify such reports as incorrect. Section 4.1 contains more details on how we handled such reports. Lastly, any vulnerabilities in the administrative interface were explicitly specified as out of scope for this study.

3.2 oDesk

oDesk is an outsourcing site that can be used to hire freelancers to perform many tasks, including web programming, development, and quality assurance. We chose oDesk because it is one of the most popular such sites, and because it gave us the most control over the hiring process; oDesk allows users to post jobs (with any specifications, payments, and requirements), send messages to users, interview candidates, and hire multiple people for the same job [1]. We used oDesk to publicize our study, hire developers that met our requirements, and pay our subjects for their work.

3.3 Subject Population and Selection

We recruited subjects for our experiment by posting our job on oDesk. We specified that respondents needed to be experienced in developing PHP applications in order to comprehend and work with our codebase, and they should have basic web security knowledge. We screened all applicants by asking them about how many times they have previously conducted a code review, a security code review, a code review of a web application, and a security code review of a web application. We also asked four multiple-choice quiz questions to test their knowledge of PHP and security. Each question showed a short snippet of code and asked whether the code was vulnerable, and if so, what kind of vulnerability it had; there were six answer choices to select from. We accepted all respondents who scored 25% or higher on the screening test. This threshold was chosen because it allowed us to have a larger sample size, while still ensuring some minimum level of knowledge and understanding of security issues.

3.4 Task

We gave participants directions on how to proceed with the code review, an example vulnerability report, and the TestCMS codebase. The instructions specified that no automated code review tools should be used. Also, the developers were told to spend 12 hours on this task; this number was calculated based upon a baseline of 250 lines of code per hour, as suggested by OWASP [2]. We designed a template that participants were instructed to use to report each vulnerability. The template has the following sections for the developer to fill in accordingly:

1. Vulnerability Type
2. Vulnerability Location
3. Vulnerability Description
4. Impact
5. Steps to Exploit

The type and location gave us basic information about the vulnerability. The template included "Vulnerability Description" and "Impact" sections in order to deter developers from using automated tools; it would be more challenging to successfully fill out these sections if a tool was used as opposed to a manual review. The last section, "Steps to Exploit", was intended to encourage developers to report only exploitable vulnerabilities as opposed to poor security practices in the code.

The developers were asked to review only a subset of the code given to them. In particular, we had them review everything but the administrative interface and the client-side code. They reviewed approximately 3500 lines of code in total. We specified our interest only in exploitable vulnerabilities. In return, we paid them $20/hour for a total of $240 for the completed job. This fixed rate leaves the relationship between compensation and reviewer effectiveness an area for future work.

3.5 Data Analysis Approach

Before the study, we scoured public vulnerability databases and Anchor's bug tracker to identify all known vulnerabilities in TestCMS. This allowed us to identify a "ground truth" enumeration of vulnerabilities, independent of those the reviewers were able to find. We manually analyzed each participant's report and evaluated the accuracy and correctness of all bugs they reported, which we describe in more detail in Section 4.1. After running statistical tests on the data, we were able to quantify how well developers conducted their reviews.

3.6 Threats to Validity

oDesk Population. As stated previously, we hired developers through the oDesk outsourcing website. This limits our population to registered oDesk users, as opposed to the population of all web developers or security reviewers. If the population of oDesk users differs significantly from the population of all web developers or security reviewers, then our results will not necessarily generalize to this larger population. However, given the success of oDesk, the population we study is interesting in its own right, and, at the very least, relevant to anyone hiring security reviewers using oDesk.

Artificial Vulnerabilities. As mentioned in Section 3.1, we introduced two SQL Injection vulnerabilities. Adding these artificial vulnerabilities creates an artificially flawed codebase where the application's original developer did not introduce all of the bugs. These artificial vulnerabilities could bias the results since it may make the code review easier or harder than reviewing a "naturally buggy" application. The changes made to the codebase were modeled after vulnerabilities found in other CMSs, which we hope will serve to minimize the artificiality of the codebase by ensuring that real web developers made the same mistakes before.

Static Analysis Tools. Using an outsourcing website to conduct our experiment restricted our ways of verifying how developers performed the review. This was considered when designing the experiment; we required specific and detailed information about each vulnerability reported. Requiring this information forced the developer to fully understand the vulnerability and how it affects the application, thereby minimizing the possibility that a reviewer would use static analysis tools and maximizing our ability to detect the use of static analysis tools. While there was no guarantee that developers completed their reviews manually, we assumed that all vulnerability reports completed sufficiently and accurately were a result of a manual code review and consequently we included them in our results.

Security Experts vs. Web Developers. We hired 30 reviewers for this study, some of whom specialized in security while others were purely web developers. Despite the use of a screening test, it is possible that web developers guessed correctly or that the screening questions asked about vulnerabilities that were significantly easier to detect than those found in a real application. In this case, web developers would be at a disadvantage. Security experts have a better understanding of the attacks, how they work, and how attackers can use them. It is possible that our screening process was too lenient and caused us to include web developers who are not security experts and would not get hired in the wild for security review. This could bias our results when measuring variability and effectiveness, but we anticipate that people hired to perform a security review in the wild may also include web developers who are not specialists in security.

Difficulties of Anchor Code and Time Frame. We found that the design of Anchor's source code made it particularly challenging to understand. It is neither well-documented nor well-structured. With the 12 hours that the reviewers had, results might not be the same as if the code were better-designed. The developers may not find all the vulnerabilities or they may give us many false positives. We have no way to verify that each person we hired spent the full 12 hours requested on the security review task, or that this was an accurate reflection of the amount of time they would spend on a real-world security review. If either of these conditions fails to hold, it could limit the applicability of our results.

Upshot. All empirical studies have limitations, and ours is no exception. Our hope, however, is that this study will encourage future studies with fewer or different limitations and that ultimately, multiple different but related empirical studies will, when taken together, give us a better understanding of the questions we study.

4   Results

4.1   Vulnerability Report Classification

In order to understand the developers' code reviews and provide results, we first had to classify each vulnerability report submitted. We place each vulnerability report into one of four categories: valid vulnerability, invalid vulnerability, weakness, or out of scope. A valid vulnerability corresponds to an exploitable vulnerability in the web application. If a developer identified one of the vulnerabilities we knew about or her steps to exploit the reported vulnerability worked, then this would be classified as valid. We made no judgments on how well she described the vulnerability. Additionally, if the developer identified a known vulnerability, but her steps to exploit it were incorrect, it was still classified as valid. When tallying the correctly reported vulnerabilities, each valid vulnerability added to this tally.

An invalid vulnerability is a report that specifies some code as being vulnerable when in fact it is not. We verified that an invalid vulnerability was invalid by attempting to exploit the application using the steps the developer provided, as well as looking in the source code. When tallying the number of falsely reported vulnerabilities, each invalid vulnerability added to the tally. Both valid and invalid vulnerabilities were included in the tally of total reported vulnerabilities.

The weakness category describes reports that could potentially be security concerns, but are either not exploitable vulnerabilities or are not specific to this web application; for example, any reports of a denial of service attack or insecure password choice are considered weaknesses. Weaknesses contributed to the total number of reports each developer submitted, but were considered neither correct nor false. Reports involving vulnerabilities in the administrative interface were placed in the out of scope category, as we asked developers not to report these vulnerabilities. Reports in this category as well as duplicate reports were ignored; they were not considered when counting the developer's total reports, valid reports, or invalid reports.

As we manually checked through all reports from all developers, we encountered one valid vulnerability reported by multiple reviewers that was not known to us when preparing the experiment. Therefore, we adjusted all other totals and calculations to take into account all seven known vulnerabilities (as opposed to the six vulnerabilities initially known to us). This new vulnerability is an additional Cross-Site Scripting vulnerability, raising the number of known Cross-Site Scripting vulnerabilities in TestCMS to four.

4.2   Reviewer Effectiveness

We measured the relative effectiveness of each developer by counting the total number of valid reports from that developer and calculating the fraction of reports that were valid. We looked at how many vulnerabilities were found and which specific vulnerabilities were found most and least commonly. One out of the four Cross-Site Scripting

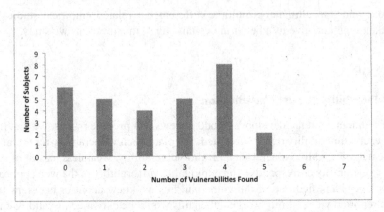

Fig. 1. The number of vulnerabilities found by individual developers

Table 1. The proportion of developers who reported each vulnerability

Cross-Site Scripting 1	37%
Cross-Site Scripting 2	73%
Cross-Site Scripting 3	20%
Cross-Site Scripting 4	30%
SQL Injection 1	37%
SQL Injection 2	20%
Cross-Site Request Forgery	17%

vulnerabilities was found by the majority of subjects, while only 17% of the reviewers found the lack of Cross-Site Request Forgery protection. Table 1 shows the percentage of developers who reported each vulnerability. The average number of correct vulnerabilities found was 2.33 with a standard deviation of 1.67. Figure 1 shows a histogram of the number of vulnerabilities found by each developer. This data shows that some reviewers are more effective than others, which addresses questions (2) and (3) in Section 2. The histogram also shows that none of the developers found more than five of the seven vulnerabilities and about 20% did not find any vulnerabilities.

4.3   Correlated Vulnerabilities

It is also interesting to note that there were cases where finding a specific vulnerability is strongly correlated to finding another specific vulnerability. Table 2 contains pairs that were significantly correlated, which is shown by their corresponding correlation coefficient (r) and p-value. The first three rows of Table 2 show vulnerabilities that were in a concentrated area in the code; this may be the reason for the correlation. The correlation between the two SQL Injection vulnerabilities may be due to the type of attack; developers may have been specifically looking for SQL Injection vulnerabilities.

Table 2. Pairs of vulnerabilities for which finding one correlates with finding the other

Pair	r	p-value
(XSS 1, XSS 2)	.4588	.0108
(XSS 1, SQLI 1)	.5694	.0010
(XSS 2, SQLI 1)	.4588	.0108
(XSS 4, SQLI 2)	.4001	.0285
(SQLI 1, SQLI 2)	.4842	.0067

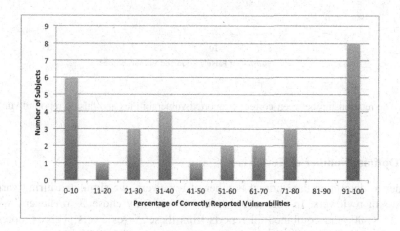

Fig. 2. The percentage of reported vulnerabilities that were correct

4.4 Reviewer Variability

While the average number of correct vulnerabilities found is relatively low, this is not indicative of the total number of vulnerabilities reported by each developer. The average number of reported vulnerabilities is 6.29 with a standard deviation of 5.87. Figure 2 shows a histogram of the fraction of vulnerabilities reported that were correct. There is a bimodal distribution, with one sizable group of reviewers having a very high false positive rate and another group with a very low false positive rate. We find significant variation in reviewer accuracy, which is relevant to questions (2) and (3) in Section 2.

Figure 3 shows the relationship between the number of correct vulnerabilities found and the number of false vulnerabilities reported. This relationship has a correlation coefficient of .3951 with a p-value of .0307. This correlation could be explained by the idea that the more closely a developer examines the code, the more possible vulnerabilities he finds, where this includes both correct vulnerabilities and false vulnerabilities. It may also reflect the level of certainty that a particular reviewer feels he must have before reporting a vulnerability. It is also somewhat similar to the trade-off in static analysis tools: when it detects fewer false alarms (false vulnerabilities), it also detects fewer true vulnerabilities; if calibrated differently, the tool may find more true vulnerabilities, at the cost of more false vulnerabilities.

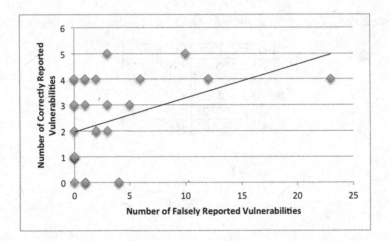

Fig. 3. The relationship between correctly reported vulnerabilities and falsely reported vulnerabilities

4.5 Optimal Number of Reviewers

In order to determine the optimal number of reviewers, we simulated hiring various numbers of reviewers. In each simulation, we randomly chose X reviewers, where $0 \leq X \leq 30$, and combined all reports from these X reviewers. This is representative of hiring X independent reviewers. For a single trial within the simulation, if this combination of reports found all seven vulnerabilities, then the trial was considered a success; if not, it was considered a failure. We conducted 1000 trials for a single simulation and counted the fraction of successes; this estimates the probability of finding all vulnerabilities with X reviewers. Figure 4 shows the probability of finding all vulnerabilities based on the number of developers hired. For example, 10 reviewers have approximately an 80% chance of finding all vulnerabilities, 15 reviewers have approximately a 95% chance of finding all vulnerabilities, and it is probably a waste of money to hire more than 20 reviewers. These results answer questions (4), (5), and (6) in Section 2.

4.6 Demographic Relationships

Our results did not indicate any correlations between self-reported demographic information and reviewer effectiveness. None of the characteristics listed in Appendix A had a statistically significant correlation with the number of correct vulnerabilities reported.

Typical hiring practices include evaluation of a candidate based on his education, experience, and certifications, but according to this data it does not have a significant impact on the effectiveness of the developer's review. We were surprised to find these criteria to be of little use in predicting developer effectiveness in our experimental data. One possible explanation for these results stems from application expectations; knowing how a system works may cause the reviewer to overlook how the system can be

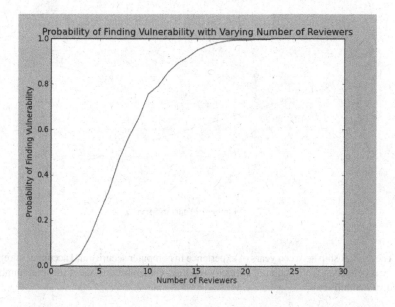

Fig. 4. A graph showing the probability of finding all vulnerabilities depending on the number of reviewers hired

used in a way that diverges from the specification [3]. Reviewers with less experience may not fully understand the system, but might be able to more readily spot deficiencies because they have no preconceived notion of what constitutes correct functionality.

Unfortunately, the design of our study did not allow us to assess the relationship between performance on the screening test and effectiveness at finding vulnerabilities, due to our anonymization procedure. We divorced the identities of participants and all information used in the hiring process from the participants' delivered results in order to eliminate possible reputation risk to subjects. We leave for future work the possibility of developing and evaluating a screening test that can predict reviewer effectiveness.

The only significant correlation found was between the number of years of experience in computer security and the accuracy of the developer's reports. We define accuracy as the fraction of correctly reported vulnerabilities out of the total reported vulnerabilities. Figure 5 shows this relationship with a correlation coefficient of -0.4141 and a p-value of .0229. While this is statistically significant in our dataset, it is not what would be expected because it indicates that the more years of experience a developer has, the lower the developer's accuracy.

We did not find significant correlations between the number of correct vulnerabilities reported and developer experience with software development, web development, or computer security. Table 3 shows the p-values for these tests.

We found a positive correlation between the number of previous web security reviews and the number of correct reports, which may be considered marginally significant. The correlation coefficient is .3117 and $p = .0936$. Figure 6 shows this relationship.

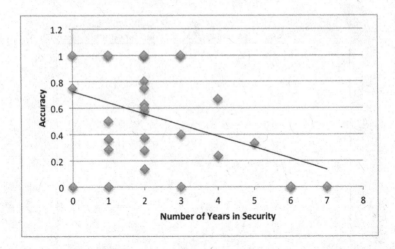

Fig. 5. The relationship between years of experience in computer security and accuracy of reports. Accuracy is defined as the fraction of correctly reported vulnerabilities out of the total number of reported vulnerabilities.

Table 3. The p-values for correlation tests between experience in different areas and the number of correct vulnerabilities reported

Area of Experience	p-value
Software development	0.6346
Web development	0.8839
Computer security	0.3612

4.7   Limitations of Statistical Analysis

A limitation of our data is that the experiment was performed with only 30 subjects. This sample size is not large enough to detect weak relationships.

5   Related Work

To our knowledge, there has been no previous research studying the effectiveness of developers at security code review. However, there have been many studies regarding the evaluation and effectiveness of code inspections. Our discussion of related work falls into three categories: code review effectiveness, methods of detecting web security vulnerabilities, and a comparison of manual code reviews to static analysis tools.

Code Review Effectiveness. One of the largest drawbacks to conducting code inspections is the time-consuming and cumbersome nature of the task. This high cost has motivated a number of studies investigating general code inspection performance and effectiveness [4–6]. Hatton [7] found a relationship between the total number of

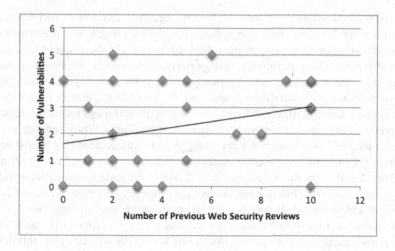

Fig. 6. The correlation between the number of previously conducted security code reviews and the number of correctly reported vulnerabilities

defects in a piece of code and the number of defects found in common by teams of inspectors. The authors gave the inspectors a program written in C with 62 lines of code; the inspectors were told to find any parts of the code that would cause the program to fail. From this research, the authors were able to predict the total number of defects. A subsequent paper by the same author [8] found that checklists had no significant effect on the effectiveness of a code inspection. Other studies have explored whether there are relationships between specific factors, such as review rate or the presence of maintainability defects, and code inspection performance [9, 10]. These experiments were carried out on general-purpose software and were not focused on security vulnerabilities, whereas our work focuses on security vulnerabilities in web applications.

Static Detection of Web Security Vulnerabilities. Most current techniques for detecting web security vulnerabilities are automated tools for static analysis. There has been work that has compared different tools and documented their differences. Bau et al. [11] showed that black-box web application vulnerability scanners do not perform well when detecting advanced and second-order forms of Cross-Site Scripting and SQL Injection. While it is more time-consuming, this may be remedied by manual code inspection. Additionally, there have been many publications evaluating and proposing new automated tools for detecting web security vulnerabilities [12–17]. Our work focuses on detecting web security vulnerabilities in a web application by manually reviewing code as opposed to using automated tools; we do not compare the reviewers' effectiveness to that of static analysis tools, though this would be a good topic for future work.

Comparing Code Review Techniques and Automated Testing. In addition to research into specific techniques for code inspection, testing, and static analysis, a number of studies have compared the effectiveness of different techniques. Basili and Selby [18] compared the effectiveness of code reading by stepwise abstraction, functional testing, and structural testing. They found that when the experiment was performed with professional programmers, code reading detected more faults than either functional or structural testing. The experiments were performed on software written in procedural languages, but none were network-facing applications. Jones [19] showed that no single method out of formal design inspection, formal code inspection, formal quality assurance, and formal testing was highly efficient in detecting and removing defects; a combination of all four methods yielded the highest efficiency. When only one method was used, the highest efficiency for removing defects was achieved by formal design inspection followed by formal code inspection. When we conducted our experiment, we did not specify how the developers should review the code as long as they did not use any automated tools. Finifter and Wagner [20] compared the effectiveness of black-box testing and manual code review for web applications, also limiting their scope to security vulnerabilities. While manual source code review was found to be more effective than automated black-box testing, black-box testing discovered vulnerabilities not found through the manual source code review.

6   Conclusion and Future Work

We designed an empirical study to ask and answer fundamental questions about the effectiveness of and variation in manual source code review for security. We hired 30 subjects to perform security code review of a web application with known vulnerabilities. The subjects analyzed the code and prepared vulnerability reports following a provided template. A post-completion survey gave us data about their personal experience in web programming and security and their confidence in their vulnerability reports.

Our results revealed that years of experience and education were not useful in predicting how well a subject was able to complete the code review. We also found that the subject's own opinion of how well they performed showed no correlation with how effective their report was. In general, we found the overall effectiveness of our reviewers to be quite low. Twenty percent of the reviewers in our sample found no true vulnerabilities, and no developer found more than five out of the seven known vulnerabilities.

No self-reported metric proved to be useful in predicting reviewer effectiveness, leaving as an open question the best way to select freelance security reviewers that one has no previous experience with. The difficulty in predicting reviewer effectiveness in advance supports anecdotal reports that the most effective way to evaluate a freelancer is by their performance on previously assigned tasks.

It would be interesting to study whether these results apply to other populations and to evaluate whether performance on our screening test correlates with reviewer effectiveness. It would also be interesting to compare the effectiveness of manual security review to that of other techniques, such as automated penetration testing tools. We leave these as open problems for future work.

Acknowledgments. We give special thanks to Aimee Tabor and the TRUST program staff. This work was supported in part by TRUST (Team for Research in Ubiquitous Secure Technology) through NSF grant CCF-0424422, by the AFOSR under MURI award FA9550-12-1-0040, and by a National Science Foundation Graduate Research Fellowship. Any opinions, findings, conclusions, or recommendations expressed here are those of the authors and do not necessarily reflect the views of the entities that provided funding.

A   Demographic and Other Factors

We tested the following demographic and other factors for correlation with number of correctly reported vulnerabilities. None were statistically significant.

- Self-reported level of understanding of the web application
- Percentage of vulnerabilities the developer thought they identified
- Years of experience in software development
- Years of experience in web development
- Years of experience in computer security
- Developer's confidence in the review
- The number of security reviews previously conducted by the developer
- The number of web security reviews previously conducted by the developer
- Self-reported level of expertise on computer security
- Self-reported level of expertise on software development
- Self-reported level of expertise on web development
- Self-reported level of expertise on web security
- Education
- Number of security-related certifications

References

1. TopSite: 10 Best Outsourcing Websites, http://www.topsite.com/best/outsourcing
2. OWASP Foundation: Code Review Metrics (2010),
   https://www.owasp.org/index.php/Code_Review_Metrics
3. Baca, D., Petersen, K., Carlsson, B., Lundberg, L.: Static code analysis to detect software security vulnerabilities—does experience matter? In: International Conference on Availability, Reliability and Security, ARES 2009, pp. 804–810. IEEE (2009)
4. Fagan, M.E.: Design and Code Inspections to Reduce Errors in Program Development. IBM Systems Journal 15(3), 182–211 (1976)
5. McCarthy, P., Porter, A., Siy, H., Votta Jr., L.G.: An Experiment to Assess Cost-Benefits of Inspection Meetings and Their Alternatives: A Pilot Study. In: Proceedings of the 3rd International Software Metrics Symposium, pp. 100–111 (March 1996)
6. Biffl, S.: Analysis of the Impact of Reading Technique and Inspector Capability on Individual Inspection Performance. In: Proceedings of the Seventh Asia-Pacific Software Engineering Conference (APSEC), pp. 136–145 (2000)
7. Hatton, L.: Predicting the Total Number of Faults Using Parallel Code Inspections (May 2005), http://www.leshatton.org/2005/05/total-number-of-faults-using-parallel-code-inspections/
8. Hatton, L.: Testing the Value of Checklists in Code Inspections. IEEE Software 25(4), 82–88 (2008)

9. Albayrak, O., Davenport, D.: Impact of Maintainability Defects on Code Inspections. In: Proceedings of the 2010 ACM-IEEE International Symposium on Empirical Software Engineering and Measurement, pp. 50:1–50:4 (2010)
10. Ferreira, A., Machado, R., Costa, L., Silva, J., Batista, R., Paulk, M.: An Approach to Improving Software Inspections Performance. In: 2010 IEEE International Conference on Software Maintenance (ICSM), pp. 1–8 (September 2010)
11. Bau, J., Bursztein, E., Gupta, D., Mitchell, J.: State of the Art: Automated Black-Box Web Application Vulnerability Testing. In: 2010 IEEE Symposium on Security and Privacy, pp. 332–345 (May 2010)
12. Huang, Y.W., Yu, F., Hang, C., Tsai, C.H., Lee, D.T., Kuo, S.Y.: Securing Web Application Code by Static Analysis and Runtime Protection. In: Proceedings of the 13th International Conference on the World Wide Web, pp. 40–52 (2004)
13. Kals, S., Kirda, E., Kruegel, C., Jovanovic, N.: SecuBat: A Web Vulnerability Scanner. In: Proceedings of the 15th International Conference on the World Wide Web, pp. 247–256 (2006)
14. Jovanovic, N., Kruegel, C., Kirda, E.: Pixy: A Static Analysis Tool for Detecting Web Application Vulnerabilities. In: IEEE Symposium on Security and Privacy, pp. 263–268 (May 2006)
15. Wassermann, G., Su, Z.: Sound and Precise Analysis of Web Applications for Injection Vulnerabilities. In: Proceedings of the 2007 ACM SIGPLAN Conference on Programming Language Design and Implementation, pp. 32–41 (June 2007)
16. Lam, M.S., Martin, M., Livshits, B., Whaley, J.: Securing Web Applications With Static and Dynamic Information Flow Tracking. In: Proceedings of the 2008 ACM SIGPLAN Symposium on Partial Evaluation and Semantics-Based Program Manipulation, pp. 3–12 (2008)
17. Kieyzun, A., Guo, P., Jayaraman, K., Ernst, M.: Automatic Creation of SQL Injection and Cross-Site Scripting Attacks. In: 31st IEEE International Conference on Software Engineering, pp. 199–209 (May 2009)
18. Basili, V., Selby, R.: Comparing the Effectiveness of Software Testing Strategies. IEEE Transactions on Software Engineering SE-13(12), 1278–1296 (1987)
19. Jones, C.: Software Defect-Removal Efficiency. IEEE Computer 29(4), 94–95 (1996)
20. Finifter, M., Wagner, D.: Exploring the Relationship Between Web Application Development Tools and Security. In: Proceedings of the 2nd USENIX Conference on Web Application Development. USENIX (June 2011)

Eliminating SQL Injection and Cross Site Scripting Using Aspect Oriented Programming

Bojan Simic and James Walden

Department of Computer Science
Northern Kentucky University
Highland Heights, KY

Abstract. Security vulnerabilities in the web applications that we use to shop, bank, and socialize online expose us to exploits that cost billions of dollars each year. This paper describes the design and implementation of AspectShield, a system designed to mitigate the most common web application vulnerabilities without requiring costly and potentially dangerous modifications to the source code of vulnerable web applications.

AspectShield uses Aspect Oriented Programming (AOP) techniques to mitigate XSS and SQL Injection vulnerabilities in Java web applications. AOP is a programming paradigm designed to address cross-cutting concerns like logging that affect many modules of a program. AspectShield uses the Fortify Source Code Analyzer to identify vulnerabilities, then generates aspects that weave in code that mitigates Cross-Site Scripting and SQL Injection vulnerabilities. At runtime, the application executes the protective aspect code to mitigate security issues when a block of vulnerable code is executed.

AspectShield was tested with three enterprise scale Java web applications. It successfully mitigated SQL Injection and Cross-Site Scripting vulnerabilities without significantly affecting performance. The use of AspectShield in these enterprise level applications shows that AOP can effectively mitigate the two top vulnerabilities of web applications in a cost and time effective manner.

Keywords: cross site scripting, xss, sql injection, SQLI, application security, aspect oriented programming, AOP, aspect, java, web application security.

1   Introduction

Most web applications contain security vulnerabilities. A recent paper shows that 71% of education, 58% of social networking, and 51% of retail websites are exposed to a serious vulnerability every day [2], and that 64% of websites have at least one information leakage vulnerability [3].

Securing web applications is an important but complicated task for any development team. While new applications can be designed with security in mind, a significant fraction of software consists of legacy applications that were not designed to be secure. This paper describes AspectShield, a system that can be applied to both new and legacy web applications to mitigate some of the most common vulnerabilities without modifying the source code of those applications.

J. Jürjens, B. Livshits, and R. Scandariato (Eds.): ESSoS 2013, LNCS 7781, pp. 213–228, 2013.
© Springer-Verlag Berlin Heidelberg 2013

In theory, legacy web applications can be rewritten to be secure. However, vulnerability remediation is expensive, with estimates of the cost of remediating a single security vulnerability ranging from $160 to $11,000 per vulnerability, depending on the type of vulnerability and its interaction with other code [22]. Web application security consultant Jeremy Grossman noted "The struggle is how do you deal with an enormous number of sites riddled with vulnerabilities? You can't just recode them. It's a dollars and cents issue."

Modification of legacy web applications also introduces the risk of altering the behavior of the application and introducing new defects [21]. Many organizations prefer to avoid modifying legacy applications where possible. Development teams are often afraid of modifying legacy applications, which unfortunately exacerbates the problem by reducing experience with the legacy application, sometimes to the point where no one who remains in the organization understands the application's design or code [21].

We designed AspectShield to mitigate vulnerabilities while avoiding the risk of altering application behavior and avoiding the cost of remediating vulnerabilities through alteration of the source code. This protection is implemented using Aspect Oriented Programming (AOP) techniques. The use of AOP allows for security logic to be developed independently of business logic. This separation of concerns produces a code base uncluttered by logging, input validation, access control, and error handling logic. While AspectShield does not modify application source code, it must have access to the source code in order to identify vulnerabilities and to recompile the application while weaving in vulnerability mitigating aspects generated by AspectShield.

We chose to use AspectJ in this paper, as it is a mature implementation of AOP, that has been in development by the Eclipse Foundation for over a decade. AspectJ is the most widely used AOP system for the Java programming language. An AspectJ aspect is composed of two major pieces:

1. The **pointcut** of an aspect is a pointer to well-defined sections of the application's source code (join points). In our system, a well-defined piece of code can be the name of a vulnerable method name with a particular signature.

2. The **advice** of an aspect defines the specific logic that is to be applied at each join point identified by a corresponding pointcut. There are three types of advice called *before*, *after*, and *around* that execute this logic before, after, or instead of the join point. The AspectShield system uses the around advice to execute validation algorithms in place of vulnerable sections of code identified by the Fortify SCA.

Aspects are woven into the byte code of the application at compile time, while the advice logic is executed at runtime at each block of code identified by pointcuts of the aspects.

The remainder of this paper is composed of the following sections. Section 2 will describe how the creation of the vulnerability mitigation aspects is accomplished. Section 3 will describe the design of an AspectShield aspect. Section 4 will go into detail about the algorithms used to mitigate SQL Injection and XSS attacks. Section 5 will provide the validation and results of an AspectShield implementation on several open source projects. Sections 6, 7, and 8 will describe related work, future work, and conclusions, respectively.

2   Generating the Security Aspects

AspectShield consists of three major steps: use of an external static analysis tool, vulnerability location based on the output of static analysis tools, and generation of aspects to mitigate the vulnerabilities and weaving of the aspects into the locations of the vulnerabilities.

2.1   Step 1 - Source Code Analysis

We first locate vulnerabilities using the Fortify Source Code Analyzer (SCA) static analysis tools. Fortify SCA is the winner of the 2011 CODiE awards for "Best Security Solution" [32] and identifies more vulnerabilities than any other detection method. The tool scans the web application source code for vulnerabilities, generating an XML report as output. Counts of vulnerabilities of each type found by Fortify SCA for the three open source web applications we used in this study are shown in Table 1 below. We ignored vulnerability reports of other types for this paper, though we plan to study them in future work.

Table 1. Fortify SCA Results

Application Analyzed	XSS Vulnerabilities	SQLI Vulnerabilities
Alfresco ECM	10	12
Apache OfBiz	869	737
JadaSite E-Commerce	11	76

2.2   Step 2 – Analyzing the SCA Results

Fortify SCA reports detailed information about vulnerabilities, including category, file location, and line number. When we analyzed the Fortify SCA reports for a number of web applications, we found that the root causes of XSS and SQL Injection vulnerabilities were a small set of functions. Functions identified as root causes include   `executeQuery()`   for   SQL   Injection   and `request.getParameter()` for XSS. We compiled a list of potentially vulnerable functions, which were stored in XML files that AspectShield uses to generate pointcuts to mitigate the vulnerabilities. The resulting XML files contained nine functions where the static analysis tool found XSS vulnerabilities and eight different definitions that correspond to potential SQL Injection vulnerabilities. For each of the functions, information such as the function name, number of parameters, and parameter types was recorded. If additional functions are discovered in the future, they can be added to the XML configuration files.

When AspectShield starts, it parses reports of XSS and SQL Injection vulnerabilities from the XML output of Fortify SCA. AspectShield asks the user to select a mitigation type for each vulnerability, which is applied by weaving in an aspect to apply that mitigation using the location information found in the SCA output.

2.3   Step 3 – Running the Aspect Generator

For each vulnerability reported by Fortify SCA, the user will be prompted to select a mitigation type. Different types of mitigations are available for XSS and SQL Injection vulnerabilities. AspectShield is designed to be used by a developer with prior experience with the application that was analyzed. As it is possible for users to select an incorrect mitigation, this user should be the person responsible for the security of the application and have training in security coding and best practices. Unfortunately, there is no universal input validation or encoding technique that could be applied to all vulnerabilities, so AspectShield must ask the user for assistance.

For XSS vulnerabilities, an AspectShield user will be provided with a list of 14 options that range from various types of encoding to whitelisting. These options are implemented using the OWASP Enterprise Security Application Programming Interface (ESAPI) library [9]. ESAPI is a free, open source library of security controls that is widely used by organizations ranging from American Express to the World Bank. It is BSD licensed, enabling AspectShield to use it without introducing licensing issues for commercial software. ESAPI features that we use to mitigate XSS include JavaScript, CSS, HTML, and other types of encoding, along with whitelist rulesets for validating data types such as email addresses and alphanumeric data. AspectShield also allows the user to provide a custom regular expression for validating input, since no library can anticipate every data type accepted by web applications.

For SQL Injection vulnerabilities, a user will be provided with the option to encode the SQL query for either the Oracle or MySQL dialects of SQL. This limitation arises from the fact that the ESAPI library only supports these two dialects of SQL. While encoding is the only option available to the user for mitigating SQL Injection, AspectShield implements additional measures to prevent exploitation of SQL Injection vulnerabilities. These measures include the removal of multiple queries, tautology detection, and the removal of SQL comments before a SQL query is executed.

Once the user selects the mitigations to implement for each vulnerability, AspectShield uses its pointcut and advice templates to generate two aspects for XSS and SQL Injection mitigation. Each selected mitigation is written to a map that is defined and populated in the corresponding aspect's constructor. The advice logic identified in the advice template will then reference this map to determine which type of mitigation should be applied based on the location of the join point.

Once the aspects have been generated, the application is ready to be recompiled with AspectJ to weave in the aspects. AspectShield provides a static JAR file containing the mitigation algorithms that will be linked into the application during recompilation. The implementation of these algorithms is defined in the following section.

3   AspectShield Design

In order to generate the SQL Injection and XSS mitigation aspects, we created templates for the pointcut and corresponding advice for each of the vulnerable methods that are intercepted to mitigate vulnerabilities.

The pointcut template contains placeholders for the method name, the method signature, the pointcut designator, and a within string. All of the pointcuts in AspectShield use the "call" designator, which allows a method to be intercepted whenever it is called. The within string placeholder will be replaced with a list of names of files in which the method should be intercepted. The method name and parameters will be retrieved from AspectShield's XML configuration files describing potentially vulnerable functions. With the pointcut template created, it will be used to create join points for all of the pieces of code where vulnerabilities were reported.

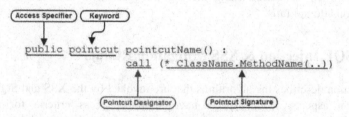

Fig. 1. Example of an AspectJ pointcut

An aspect's advice will be executed at every join point matched by the pointcut in the application's source code. The around advice used in this implementation executes code in place of the join point it operates over. Since it can have a return value, it must be given a return type (Figure 2).

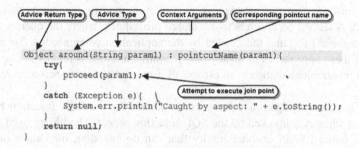

Fig. 2. Example of an AspectJ Advice

Inside of the around advice, the original join point can be executed using the proceed function which takes the same arguments as the join point. Much like the pointcut template defined above, the advice template contains placeholders that are populated when AspectShield is executed. The advice will have a corresponding name equal to the pointcut that it will execute upon. Depending on whether the aspect being created is for XSS or SQL Injection mitigation, the advice will make a call to the appropriate validation algorithm that will do the mitigation. The algorithm will be provided the original, potentially malicious, parameters for each join point and will return a safe value.

The advice also contains logic to determine whether or not the user elected to provide mitigation for a particular join point. In the event that the call to the mitigation algorithm fails, the advice will execute its `proceed()` method with the pointcut's default parameters in order to maintain the application's normal execution flow. This ensures that AspectShield will not break any of the application's functionality. AspectShield's first priority is to maintain application functionality even in the unlikely event that its mitigation algorithm fails, as security fixes should not break the application. However, if desired, the tool could easily be modified to prevent the code from executing in this scenario. The advice also logs each mitigation using the log4j logger, so a user can detect when a mitigation attempt fails.

4   SQL Injection & XSS Mitigation Results

This section describes the algorithms that are invoked by the XSS and SQL Injection mitigation aspects. Both aspects have similar success criteria for application performance, correct execution of application code, and vulnerability mitigation. When an AspectShield aspect is invoked at runtime, it will receive the potentially malicious input and the mitigation type to be applied as parameters.

4.1   The SQL Injection Mitigation Algorithm

There are three primary choices of mitigation technique for SQL injection vulnerabilities. The first is to use parameterized queries or prepared statements. This method ensures that the attacker is not able to modify the query that is being executed. A second approach is to use stored procedures, where the queries are stored in the database itself and then called by the application when needed. To implement either of these approaches in legacy code, significant work is required. The approach taken in this implementation is to escape all user supplied input before executing any query.

SQL Injection mitigation will be accomplished by the SQL Injection mitigation algorithm when it is invoked by the SQL Injection aspect. The library used to encode all input is the ESAPI encoder library that can do encoding for Oracle or MySQL dialects. The steps of the SQL Injection mitigation algorithm are:

1. The SQL Injection aspect generated in the previous section will call its `doSQLInjectionFix()` method, passing it the query that needs to be validated and the encoding type specified when the user ran AspectShield to generate the aspects.
2. The validator will then test the query for any comments and remove them if found. The query will be passed to the JSQL Parser library that will parse the query and return a list of expressions that the query contains.
3. Each expression in the query will be encoded using either the MySQL or Oracle encoder depending on the choice made by the user when AspectShield was run.
4. Each expression will be tested to determine if it is a tautology, as SQL Injection exploits frequently use tautologies while normal SQL queries do not. This is done by using the Java ScriptEngineManager's Javascript engine to evaluate the expression's value. If the result is always true, the expression is marked as a tautology and removed from the original query.
5. Once all expressions are encoded and tautologies removed, the query is reconstructed using the safe values and returned to the SQL Injection mitigation aspect.
6. When the aspect receives the newly safe version of the query, it will invoke it's proceed() method and pass it the new, safe value.

The mitigation algorithm was timed at each step and performance was evaluated for three case study projects. In the event that the algorithm fails due to an inability to parse the query or for any other reason, it will catch any exceptions, log the failure using a logj4 logger, and return the original query passed in. The original query passed into the algorithm is returned so that if the algorithm fails, the application's normal execution flow will not be affected.

4.2   The XSS Mitigation Algorithm

The difficulty in preventing XSS comes from the fact that such a large number of attack vectors exists. An attacker could potentially steal the session of a victim, manipulate files on the victim's computer, record all keystrokes the victim makes in a web application, or probe a company's intranet where the victim is located [52]. Appropriate validation and encoding can address most reflected and stored XSS vulnerabilities. The algorithms described in this section use the ESAPI encoder and validator libraries to perform escaping and encoding of dangerous data.

The XSS mitigation aspect contains pointcuts that intercept functions, such as getparameter() from the request object and println() that were identified by the Fortify SCA. At each join point, the aspect's advice logic will implement either encoding or whitelisting on the value intercepted by each pointcut. The process is outlined in the steps below:

1. The advice will call a doXSSFix() method for each join point. The method will be passed the intercepted parameter, as well as the type of fix to implement as chosen by the user during the aspect generation phase.

2. Depending on the type of fix the user selected for the join point, the XSS Validator will apply either encoding for the chosen format or validation using a whitelist. The user has the option to choose from several types of encoding or whitelist provided by the ESAPI library [53] [54].
3. If the desired mitigation is a whitelist, the algorithm will check the input against a particular regular expression. If the input fails to match, the code will remove any characters that do not match the desired character set.
4. If the desired mitigation is a particular type of encoding (CSS, JavaScript, HTML, etc…), the ESAPI library will be used to encode the input.
5. The last step is for the XSS Validator to return the resulting string back to the aspect's advice. The advice will then call the proceed() method and pass it the encoded string.

The most difficult aspect of implementing the XSS validator algorithm was to catch all possible exceptions that the ESAPI encoder and validator classes can throw. In the event that an exception occurs, AspectShield handles it gracefully to ensure that no functionality of the web application is broken. The algorithm also supports different types of input such as String and byte arrays in order to support all possible join points identified by the SCA.

5   Validation and Results

This section describes the evaluation of the work. The evaluation will show that AspectShield successfully mitigates both SQL Injection and XSS vulnerabilities without altering source code or breaking application functionality. The evaluation will also show that the libraries and algorithms used to eliminate two of the most important web application vulnerabilities are not only functional but also do not impact application performance by more than an average of 1.99ms per request. Both of the aspects generated by this program will be evaluated in a live environment because they will be built into and executed as part of each of the three case study applications chosen.

5.1   Identifying the Case Study Applications

Since we used AspectJ for AspectShield, our case studies are web applications written in Java. Our other selection criteria for applications included availability of source code, application size of at least 300 classes, support of MySQL or Oracle databases, and developer activity. Websites such as FreshMeat.net, SourceForge.net, and GitHub.com were searched to find suitable candidates.

The first project selected was a popular open source enterprise content management framework called Alfresco. This program has over 140,000 community members, over 2000 enterprise customers, and over 3,000,000 downloads [43]. This application was chosen because as a content management application, it has many points of entry that could potentially be exploited by hackers.

The second project chosen was Apache's OfBiz. This application is one of the Apache Software Foundation's projects and is one of the best open source ERP and E-Commerce implementations. OfBiz was chosen because of its use in E-Commerce, where these types of applications are heavily targeted because they contain personal information such as addresses and credit card information.

The last project chosen was a less well known application called JadaSite, which is another open source E-Commerce framework. This project was chosen because it makes heavy use of newer web technologies such as AJAX and WYSIWYG user interfaces.

5.2   Methods for Validating AspectShield

We perform three types of validation for AspectShield. First, we evaluate the algorithms and libraries invoked by aspects at each join point. Second, each of the aspects will be evaluated as part of the Alfresco, OfBiz, and JadaSite applications. The first method for evaluation is to determine the functionality of the utility class the aspects call at each join point, the XSS Validator library, and the SQL Injection Validator library. Third, we create unit tests for each of the libraries' methods with JUnit4, and then running a stress test to measure the performance of the libraries.

5.3   SQL Mitigation Algorithm JUnit Results

For the initial set of tests, a list of twenty-one SQL queries was executed fifty times for both the ESAPI MySQL and Oracle encoders. None of the queries exceeded 300 characters. The list contained different types of malicious content that could result in exploits ranging from injection of scripts to bypassing login forms. After the JUnit4 test was executed, the log file that contained the results of each query test was analyzed. The result showed several promising indicators that the desired results were achieved in both successfully mitigating SQL Injection and doing so in less than 5ms on average. The results of the log file provided information such as removal of comments from the query and tautology detection and removal. One such example is on the SQL query "SELECT * FROM Users WHERE ((Username='1' or '1' = '1'))/*') AND (Password=MD5('password')))". This query contains a comment that would bypass the password checking for login validation as well as a tautology, '1' = '1', that would also bypass login checks. This query was successfully mitigated by removing the comment entirely and replacing the tautology with an expression that would evaluate to false.

The second characteristic of SQL Injection Validator execution that was analyzed was the execution time for each of the query validations (Figure 3). Data was collected from the log file, then the maximum, minimum, average, median, and mode were all calculated. The longest execution time was 33ms, and the shortest was 1ms. The average query validation time was only 5.79ms, and the most common time, the mode, was 5ms. The fact that the longest execution period was only 33ms was very encouraging considering that it was much shorter than the average request.

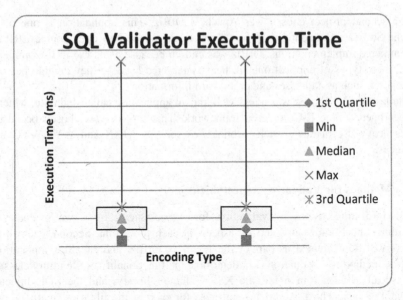

Fig. 3. SQL Mitigation Algorithm execution time

5.4   XSS Mitigation Algorithm JUnit Results

The evaluation results for testing the XSS Validator contained a considerably larger amount of information because of the many different mitigations that are available to the user. However, the 13 different categories can be split into two separate categories that had similar results. The "encoding" category consists of mitigations that use the ESAPI encoder library, which simply encodes the input characters according the context in which they would be used. The second and generally more involved category is "whitelist", where user input is compared to a regular expression string. The results for both can be found in the table below:

Validation Type	Average (ms)	Max (ms)	Min (ms)	Median (ms)
HTML Attribute Encode	0.04	13	0	0
Email Whitelist	3.1	197	0	1
Alpha Numeric Whitelist	3.3	199	0	1
URL Encoding	0.09	6	0	0
SSN Whitelist	3.06	199	0	1
Zip Code Whitelist	3.22	196	0	1
Credit Card Validation	4.53	229	0	1
HTML Encoding	0.05	11	0	0
CSS Encoding	0.03	1	0	0
Alpha Whitelist	3.3	209	0	1
Javascript Encoding	0.08	47	0	0
IP Address Whitelist	3.08	198	0	1

As shown above, the items in the "encoding" category had significantly lower average and maximum execution times. The lowest of these was for CSS encoding, which only took a third of a millisecond, and the highest was 47ms for the JavaScript encoding of a string. The longest execution time belonged to credit card validation, which took almost a quarter of a second at 229ms. The reason that this function takes so long to execute is because the ESAPI validator evaluates the string to see if it matches several different credit card patterns and computes the Luhn checksum, which is used to validate credit card numbers.

5.5   Results of Integration with Case Study Applications

The final part of the evaluation process was to test the aspects with three case study applications running in a live environment. The difficulty of this validation step comes from the fact that the vulnerabilities identified by the SCA are scattered throughout the application and the path to each of these can be difficult to reach since all the projects are enterprise scale applications. Therefore, the chosen approach is to only evaluate the execution of join points that are easily reachable by a typical user of the application. When doing this step, the aspect generation program must either be extracted into a JAR or included in the build path of each of the applications so that the XSS and SQL Injection Validators as well as required libraries such as the JSQL Parser can be referenced.

For the OfBiz project, 22 join points were evaluated and the execution time and result were analyzed. Each join point was able to execute successfully and with similar time to the JUnit4 execution data explained above. Out of the 22 join points evaluated, 16 were possible XSS vulnerabilities with whitelist validation and 6 applied SQL encoding. Since the Alfresco project consists of multiple projects with different contexts, the "repository" project was chosen for execution since it contained a fairly large portion of the vulnerabilities identified. This project contained 11 SQL Injection and 4 XSS vulnerabilities. Both of the aspects executed as expected, except one in the SQL Injection category, where a table creation statement was not supported by the JSQL Parser. However, even though the query was not supported, the program still executed normally, because the SQL Injection Validator returns the original value if query validation fails. The JadaSite project had no issue executing any of the 8 XSS and 14 SQL Injection join points that were tested.

5.6   Evaluation – OWASP Webgoat Project

To provide an extra layer of validation using an application with which most professionals in the web application security field are familiar, AspectShield was applied to the OWASP WebGoat version 5.2 project. WebGoat is a deliberately insecure Java web application that is designed to teach web application security. It contains a number of purposefully implemented vulnerabilities, including several SQL Injection and XSS vulnerabilities.

Fortify SCA was used to locate these vulnerabilities. Using the Aspect Generator, aspects were created to implement mitigations at runtime by intercepting potentially malicious user input. The application was compiled and deployed with the generated aspects, and then each of the XSS and SQL Injection modules were tested.

The two modules tested manually were Injection flaws and XSS. For SQL Injection, several types of queries such as insert, update, and delete were executed as well as inputs with multiples queries and tautologies. All of the queries attempted to execute would normally exploit the application but were successfully mitigated with AspectShield. For XSS vulnerabilities, the fixes specified to the XSS aspect during the generation of the aspects were to do Javascript encoding on various inputs. The aspect successfully mitigated reflected, stored, and DOM based XSS attacks that would normally succeed in the modules associated with that type of vulnerability.

5.7   Evaluation Conclusion

Evaluation showed that AspectShield successfully prevented the exploitation of XSS and SQL INJECTION vulnerabilities in three case study Java web applications, as well as in the OWASP WebGoat application. JUnit testing of both the SQL INJECTION and XSS mitigation algorithms proved that the implementation would not significantly affect application performance or interrupt execution flow. The implementation of the vulnerability mitigation aspects across three enterprise level web applications showed that the implementation can easily be applied to existing code and successfully mitigate attacks.

6   Related Works

6.1   AOP and Security

Security with AOP has been the subject of study in several different publications [6], [7], [8]. Some of the papers that influenced this work include the work done by Robin C. Laney and Janet van der Linden [7], where the authors were able to leverage the power of AOP in order to make significant changes to legacy applications. This was particularly interesting, because often programmers are assigned the task to implement some improvement to a piece of software that has not been modified for several years and has little documentation. The authors showed that programmers can use AOP to evolve legacy code and leave behind digital signatures that reduce the likelihood of breaking existing functionality while enhancing the application overall. In the work done by Minhuan Huang, Lufeng Zhang, and Chunlei Wang [8], they created a fully functional library that implements security features across an application using AOP. Although their library mostly focused on encryption and decryption, it showed how security could be implemented using AOP.

Seinturier and Hermosillo wrote a paper which relates closely to this work [9]. Their tool, AProSec, detects inputs to a web application using aspects to intercept potential XSS and SQL Injection attacks. Their aspects then either warn the user or reject the potentially harmful data input. Their approach was unique in that they wrote aspects that implement security detection functionality in a web application server's native libraries without having to modify the web application server code or write their own. Some of their concepts such as intercepting request and response parameters were leveraged in the creation of aspects in this research. Another framework created by Zhi Jian Zhu and Mohammad Zulkernine uses AOP for intrusion detection for some of the most common attacks in the Web Application Security Consortium [23]. The AspectShield tool takes this approach in a different direction by intercepting vulnerable code detected by the SCA and than fixing harmful inputs during run time.

While we were making final revisions of this paper for publication, we discovered a paper describing the implementation of a system similar to AspectShield that had been published after we completed the version of this paper that we submitted for publication. This paper describes the creation of an Eclipse IDE plug-in that does automatic discovery of weaknesses in the application code and with the assistance of the developer, remediates them with AOP [24]. The plug in also makes use of the ESAPI libraries created by OWASP and generates aspects based off of user selected validation and encoding techniques.

6.2   Security with AOP for SQL Injection

When looking at the most prevalent web application security vulnerabilities, injection is typically at the top of the list at every reliable source. SQL Injection tends to be the most harmful of these and it is ranked as the second most common form of attack on web applications [10]. One of the most extensive works of research done by V. Shanmughaneethi, Yagna Pravin, and Emilin Shyni uses aspects to analyze a SQL query for potentially malicious content [10]. This tool uses aspects which call web services to analyze queries and create errors in order to prevent malicious SQL from being executed. This is a good approach in theory, but the authors do not specifically discuss the implications of making web service calls with respect to performance and reliability of these web services.

In his book [11], Justin Clarke briefly discusses how AOP can be leveraged to hot-patch applications that are vulnerable to SQL Injection at runtime. He recommends using one of the AOP implementations such as AspectJ and Spring AOP to implement checks for insecure dynamic SQL libraries. Most of the references, such as Clarke's book, offer a few sentences on how the paradigm could be used but do not reference any concrete implementations. Even solutions that do provide concrete implementations, such as the Shanmughaneethi paper, only work as far as identifying vulnerabilities but don't do much to mitigate them.

6.3   Security with AOP for XSS

AOP can be used to mitigate and in some cases eliminate XSS vulnerabilities in web applications. This is especially true when the application in question would require a complete re-write in order to achieve security [23]. According to OWASP, the best two ways to prevent XSS is to escape all untrusted data based on the content of the web page and to do whitelist validation on user inputs [21]. Using AOP, a developer can create aspects to intercept incoming and outgoing data that would be displayed to the user and apply either escaping or whitelisting without modifying the existing source code. Mece and Kodra [24] were able to create a XSS validation aspect that does whitelisting of user inputs. Their "validator" aspect treated all strings that were not alphanumeric as potentially dangerous and denied them. While this is certainly a very safe approach, applying such restrictions onto an existing application would almost certainly break functionality because many applications require inputs much more complicated than just an alphanumeric string.

There have been multiple studies with the intent to use AOP to eliminate XSS vulnerabilities [28] [29] [30]. However, most of these simply discuss the idea of using aspects in order to achieve security and very few have working implementations. Of the papers with implementations, the implementations are simple ones such as regular expression whitelisting of input that just demonstrate the potential usefulness of the AOP paradigm without offering in-depth solutions.

7   Future Works

There are a number of features and improvements that can be added to AspectShield to make it a more effective security tool and provide a better overall user experience. The first is to extend the program to support mitigation of additional types of vulnerabilities. We intend to examine the possibility of adding additional mitigations for the remaining vulnerabilities of the OWASP Top 10. Along with support for mitigation of a wider range of threats, it would also be helpful to create a graphical user interface for the aspect generator program in order to improve user experience.

A second potential area of improvement is to extend AspectShield to other programming languages. One challenge will be finding a suitable AOP implementation for each additional programming language to be supported. Since the implementation for each language would be different, new template files, function definitions, and libraries for mitigating vulnerabilities would need to be created for each programming language.

A third area for future work would be to extend the program to support multiple static analysis tools.   Different automated static analysis tools find different vulnerabilities in source code. Additional plans for future work include an Eclipse plug-in that does the mitigation aspect generation automatically, and a multi language API for creating all parts of the security aspect generation process.

8   Conclusions

Two of the most common vulnerabilities in web applications are SQL Injection and Cross-Site Scripting. Thousands of web applications process personally identifiable information such as SSNs, credit card numbers, and addresses every day, and many of these applications have a significant number of SQL INJECTION and XSS vulnerabilities that can be exploited by a malicious user. The AspectShield tool described in this paper creates aspects that prevent malicious content from being executed or stored in Java web applications using the results of the Fortify Source Code Analyzer and the users' choice of mitigation technique. The most significant feature of the approach identified is creation of the AspectShield tool, which does mitigation of vulnerabilities without the need to modify potentially fragile source code.

Our approach was to apply the AOP paradigm to execute validator classes at the locations of the vulnerabilities identified by Fortify SCA. This modular approach to implementing security allows the developers to use separation of concerns where they can apply any future security algorithms to a single location. When the user executes AspectShield, they will be given a choice of fixes to implement for each of the vulnerabilities detected by the static analysis tool. The two resulting aspects, one for XSS and the other for SQL Injection, contain pointcuts and advices that will isolate join points throughout the applications' source code and weave in the necessary code that will ensure the mitigation of these vulnerabilities.

In order to evaluate the success of the aspects created, WebGoat and three enterprise level open source Java web applications were chosen as case studies. These applications are from the E-Commerce, content management, ERP, and document management categories. AspectShield generated vulnerability mitigation aspects based on static analysis of these applications. Evaluation consisted of unit tests using the JUnit testing framework, integration of aspects into each project, and testing the mitigation aspects as part of the running applications. The evaluation proved that each of the aspects not only mitigate XSS and SQL INJECTION attacks but also do it very efficiently with most execution times being less than 10 milliseconds. The low execution time of the aspects' at each join point is significant because it is very important that the introduction of the security code did not heavily affect the execution time of the original applications.

In conclusion, this paper describes the implementation of a program that generates XSS and SQL INJECTION mitigation aspects that can be applied to mitigate vulnerabilities in both new and legacy web applications using information from static source code analysis. The main advantage of this approach compared to others evaluated that it does not require any modification to legacy code and provides a centralized location for the application's security logic. The evaluation of generated aspects with three enterprise level projects provides a great level of confidence that the approach is both valid and effective at mitigating some of the most prevalent threats to web application security.

References

1. Webroot. State of Internet Security – Protecting Enterprise Systems [Whitepaper]. Webroot Software Inc., USA (2007)
2. Electronista. LulzSec hacks Sony Pictures, reveals 1m passwords unguarded. Electronista Media Inc. (June 2, 2011)
3. Measuring Website Security: Windows of Exposure. WhiteHat Website Security Statistic Report (March 14, 2011)
4. Shanmughaneethi, V., Yagna Pravin, R., Emilin Shyni, C., Swamynathan, S.: SQLIVD - AOP: Preventing SQL Injection
5. OWASP (Open Source Web Application Security Project). OWASP Top 10 – 2010 Edition. OWASP Foundation (2010)
6. Fortify Source Code Analyzer – Capabilities. HP Fortify. Web (2011)
7. Laddad, R.: AOP @ Work: AOP Myths & Realities. IBM Developer Works (February 14, 2006)
8. ESAPI Interface Encoder. The Open Web Application Security Project. Web (2011)

 9. ESAPI Validator Library. The Open Web Application Security Project. Web (2011)
10. Li, S.: AOP: Patching in the 21st Century. Developer Fusion. Web (July 23, 2010)
11. Bostrom, G.: Database Encryption as an Aspect. In: Proceedings of AOSD 2004 Workshop on AOSD Technology for Application level Security (March 2004)
12. Laney, R., van der Linden, J., Thomas, P.: Evolution of Aspects for Legacy System Security Concerns. In: Proceedings of AOSD 2004 Workshop on AOSD Technology for Application level Security (March 2004)
13. Huang, M., Wang, C., Zhang, L.: Toward a Reusable and Generic Security Aspect Library. In: Proceedings of AOSD 2004 Workshop on AOSD Technology for Application level Security (March 2004)
14. Hermosillo, G., Gomez, R., Seinturier, L., Duchien, L.: Using Aspect Programming to Secure Web Applications. Journal of Software 2(6) (December 2007)
15. Clarke, J.: SQL Injection Attacks and Defense, 1st edn. Syngress (May 13, 2009) (March 1, 2011)
16. Mece, E., Kodra, L.: Towards full protection of Web Applications based on Aspect Oriented Programming. GJCST, 33–37 (2012)
17. Arthur, C.: Twitter users including Sarah Brown hit by malicious hacker attack. Guardian News (September 21, 2010)
18. Win, B., Shah, V., Joosen, W., Bodkin, R. (eds.): AOSDSEC: AOSD Technology for Application-Level Security (March 2004)
19. Bodkin, R.: Enterprise Security Aspects. In: Win, B., Shah, V., Joosen, W., Bodkin, R. (eds.) AOSDSEC: AOSD Technology for Application-Level Security (March 2004)
20. Fortify. Leading Bank Turns Security into a Differentiator with Fortify SCA. Fortify Software Inc. (2008)
21. Feathers, M.: Working Effectively with Legacy Code. Prentice Hall (2004)
22. Higgins, K.J.: The Cost of Fixing an Application Vulnerability. Security Dark Reading (May 11, 2009), http://www.darkreading.com/security/news/
23. Zhu, Z.J., Zulkernine, M.: A model-based aspect-oriented framework for building intrusion-aware software systems. Information and Software Technology 51(5), 865–875 (2009)
24. Serme, G., De Oliveira, A.S., Guarnieri, M., El Khoury, P.: Towards Assisted Remediation of Security Vulnerabilities. In: 6th International Conference on Emerging Security Information, Systems and Technologies (August 2012)

Author Index